MCQ's IN AGRICULTURAL MICROBIOLOGY

Editors :

Dr. Deepak G. Panpatte

Dr. Yogeshvari K. Jhala

Dr. Rajababu V. Vyas

Smt. Harsha N. Shelat

SCIENTIFIC PUBLISHERS

Published by

SCIENTIFIC PUBLISHERS (INDIA)
5 A, New Pali Road, P.O. Box 91
JODHPUR 342 001 (India)
E-mail: info@scientificpub.com
Website: www.scientificpub.com

ISBN : 978-93-89184-55-6
978-93-89184-56-3 [E/B]

Printed in India.

PREFACE

Majority of the entrance examinations for higher studies in agricultural sciences are based on the Multiple choice questions. The book is primarily meant for students appearing for UG and PG level entrance examination like Indian Council of Agricultural Research (ICAR) – AIEEA UG, AIEEA PG, AICE SRF/JRF (PGS), ICAR NET, CSIR NET, ARS prelim, PG entrance of SAUs and as well as UGC NET examination which contains Agricultural Microbiology as subject in the syllabus of the examinations. The book comprise of all the basic and applied aspects of Microbiology starting with history of microbiology and end up with microbial biotechnology.

This book serves as ready reckoner for Multiple Choice Questions in Microbiology/ Agricultural Microbiology. The book comprise of multiple choice question which are drawn by referring majority of the text books of Agricultural Microbiology. This will save valuable time and energy of readers. This book is particularly intended to match any standard agricultural microbiology textbook and to provide the students with a feedback on their improvement and prospect to get better. The book not only serves as ready means of self-assessment but also to provide the knowledge of the subject in tutorial form which is expected to provide starting point for the further studies and dialogue. Both the teachers and students of agricultural microbiology will find this book useful. A quick outlook of the questions will offer verification that the book aim to encourage way of thinking.

— Deepak G. Panpatte

Dedicated to

My Beloved Parents

&

Poornawad Vardhishnu P. P.
Dr. Vishnu Ramchandra Parnerkar

ACKNOWLEDGEMENTS

There remains the pleasant responsibility of acknowledging my gratitude and thankfulness to the team of agricultural microbiologists of Department of Agricultural Microbiology, B. A. College of Agriculture, Anand Agricultural University, Gujarat, India, who have kindheartedly supported me to bring out this valuable book for the students of agricultural science. I wish particularly to express my warm gratitude to the Hon. Vice-chancellor, Director of Research & Dean PG Studies and Principal & Dean (Agri.) of Anand Agricultural University for supporting me to compile the volume. Last but not the least; my thanks are due to my family members for their continual encouragement.

— Dr. Deepak G. Panpatte

CONTENT

Chapter 1

HISTORY OF MICROBIOLOGY

1. **Theory of abiogenesis was proposed by ....**
   (A) Aristotle (B) Robert Hook
   (C) Robert Koch (D) Louis Pasteur
2. **Who devised three jar experiments for disproving spontaneous generation theory?**
   (A) John Needham
   (B) Louis Pasteur
   (C) Francesco Redi
   (D) Theodore Schwann
3. **Who invented the microscope in 1666?**
   (A) Robert Hook
   (B) A. V. Leeuwenhoek
   (C) Robert Koch
   (D) Louis Pasteur
4. **Who first time used heat for passing air in to the growth medium?**
   (A) Girolamo Fracastoro
   (B) Theodor Schwann
   (C) Agostino Bassi
   (D) Robert Koch
5. **Who introduced the technique of cotton plugging?**
   (A) Joseph Lister
   (B) Joseph Meister
   (C) John Needham
   (D) Friedrich Schroder and Theodor Von Dusch
6. **Who devised swan neck flask experiment to disprove spontaneous generation theory?**
   (A) Felix Pouchet (B) John Tyndall
   (C) Louis Pasteur (D) Robert Koch
7. **Magnification power of Leeuwenhoek's microscope was**
   (A) 50-300 times
   (B) 100-1000 times
   (C) 10-100 times
   (D) 25-50 times
8. **Leeuwenhoek reported micro-organism as**
   (A) Aminacules (B) Animalcules
   (C) Actinomycets (D) Arthropod
9. **Leeuwenhoek first reported discovery of microorganisms to ....**
   (A) British king
   (B) Royal society of London
   (C) France royal society
   (D) Royal society of America
10. **Who have reported that invisible organisms cause disease?**
    (A) Pasteur (B) Koch
    (C) Berkeley (D) Fracastoro
11. **Who first time proved that micro-organisms cause diseases?**
    (A) Robert Koch
    (B) Louis Pasteur
    (C) Agostino Bassi
    (D) Berkeley
12. **Agostino Bassi studied which disease?**
    (A) Silkworm disease
    (B) Pig disease
    (C) Cholera
    (D) Rabies

13. Which type of single-celled organisms was first discovered by A.V. Leeuwenhoek?

(A) Bacteria (B) Fungi

(C) Protozoa (D) Nematode

14. Pasteur studied which disease of silkworm

(A) Rabies (B) Anthrax

(C) Pupal (D) Pebrine

15. Causative agent of Pebrine disease?

(A) Protozoa (B) Virus

(C) Bacteria (D) Nematode

16. Indirect evidence about association of microorganisms with human diseases was provided by

(A) Joseph Meister (B) Joseph Lister

(C) John Tyndall (D) Robert Koch

17. Joseph Lister sterilized surgical instruments with

(A) Phenol

(B) Heat and Steam

(C) Carboxylic acid

(D) Ethylene oxide

18. Joseph Lister used which chemical for sterilization of surgical instruments and dressings?

(A) Alcohol

(B) Ethylene Oxide

(C) Phenols

(D) Spirit

19. First direct evidence about involvement of microorganisms in causing disease was provided by

(A) Robert Koch (B) Louis Pasteur

(C) Camberland (D) Robert Hook

20. First direct evidence about involvement of microorganisms in causing disease was provided by studies on

(A) Rabies disease

(B) Cholera disease

(C) Anthrax disease

(D) Pebrine disease

21. Bacteria were first cultured by Robert Koch on ....

(A) Nutrient agar

(B) Cut boiled potato

(C) Boiled agar

(D) Boiled yeast

22. Solidifying agent used in culture media by Robert Koch was ....

(A) Silica gel (B) Agar

(C) Yeast extract (D) Potato

23. Gelatin melts

(A) Above 20 °C (B) Above 50 °C

(C) Above 100 °C (D) Above 28 °C

24. Who discovered agar powder for solidification of culture media?

(A) F. E. Hesse (B) Robert Koch

(C) Louis Pasteur (D) Richard Petri

25. Melting point of agar agar is

(A) > 121 (B) >100

(C) > 180 (D) < 100

26. Who discovered petri dish?

(A) Robert Koch (B) Richard Petri

(C) Fanny Hesse (D) Louis Pasteur

27. Who discovered pure culture technique?

(A) Richard Petri

(B) Louis Pasteur

(C) Antony Von Leeuwenhoek

(D) Robert Koch

28. Who first developed culture medium?

(A) Robert Koch (B) Richard Petri

(C) John Tyndall (D) Louis Pasteur

29. Who studied ammonification of organic nitrogenous substances by soil micro-organisms?

(A) J. G. Lipman and P. E. Brown

(B) M. W. Beijerinck and S. N. Winogadsky

(C) R. Russel and M. Hutchinson

(D) H. Hellriegel and H. Wilfarth

30. Who isolated tuberculosis bacteria *Mycobacterium tuberculosis*?
(A) A.V. Leeuwenhoek
(B) Louis Pasteur
(C) Robert Koch
(D) Ivanovskaya

31. Who first time prepared porcelain filter?
(A) Pierre Roux
(B) Emil von Behring
(C) Albert Whitman
(D) Charles Chamberland

32. Which finding made possible the discovery of viruses?
(A) Whatman filter
(B) Membrane filter
(C) Porcelain filter
(D) Silica filter

33. The first viral pathogen to be studied was
(A) Tobacco necrosis virus
(B) Tobacco mosaic virus
(C) Cauliflower mosaic virus
(D) Influenza virus

34. Who is known as father of bacteriological techniques?
(A) Robert Koch
(B) Louis Pasteur
(C) Elie Metchnikoff
(D) A.V. Leeuwenhoek

35. Who first discovered attenuated bacteria are not capable of causing disease?
(A) Elie Metchnikoff
(B) Robert Koch
(C) Pasteur and Roux
(D) Charles Chamberland

36. Who first discovered vaccine?
(A) Robert Koch (B) Edward Jenner
(C) Joseph Meister (D) Louis Pasteur

37. Edward Jenner in 1796 discovered vaccine
(A) Rabies (B) Hepatitis B
(C) Smallpox (D) Chicken pox

38. Who developed anthrax vaccine?
(A) Louis Pasteur
(B) Robert Koch
(C) Charles Chamberland
(D) Elie Metchnikoff

39. Who discovered rabies vaccine?
(A) Robert Koch
(B) Charles Camberland
(C) Edward Jenner
(D) Louis Pasteur

40. Attenuated anthrax vaccine was prepared by treating culture with
(A) Potassium permanganate
(B) Temperature of 42 to 43°C
(C) Sodium hypochlorite
(D) Formaldehyde

41. Attenuated rabies vaccine was prepared by growing bacteria in
(A) Rabbit (B) Mice
(C) Monkey (D) Human

42. Name the first person who has been vaccinated?
(A) Landsteiner (B) Joseph Lister
(C) Joseph Meister (D) Mark Lister

43. Who first discovered diphtheria antitoxin?
(A) Emil von Behring
(B) Landsteiner
(C) Mark Lister
(D) Robert Koch

44. Who won Nobel Prize for discovery of diphtheria antitoxin?
(A) Emil Von Behring
(B) Shibasaburo Kitasato
(C) Louis Pasteur
(D) Robert Koch

45. Antibody mediated immunity is known as
(A) Cellular immunity
(B) Innate immunity
(C) Humoral immunity
(D) Artificial immunity

46. Blood cell (leucocyte or WBC) mediated immunity is known as
(A) Humoral Immunity
(B) Cellular immunity
(C) Natural immunity
(D) Innate immunity

47. The phenomenon of engulfment of disease causing microorganisms by blood cells or leucocytes of WBC is known as
(A) Phagocytosis
(B) Immunity
(C) Leucocytosis
(D) Erythrocytosis

48. Phagocytosis was first discovered by
(A) Elie Von Behring
(B) Robert Koch
(C) Elie Metchnikoff
(D) Charles Camberland

49. The cells which engulf microorganism are known as
(A) Lymphocytes (B) Hemocytes
(C) Phagocytes (D) Endocytes

50. Pasteur discovered that alcohol become sour due to
(A) Overgrowth of yeast
(B) Lactic acid bacteria
(C) Propionic acid bacteria
(D) Slow growth of yeast

51. Which compound is responsible for souring of alcohol?
(A) Lactic acid
(B) Propionic acid
(C) High alcohol content
(D) High sugar content

52. All fermentations were due to activity of specific yeast and bacteria was proven by
(A) Louis Pasteur
(B) Rober Koch
(C) Elie Metchnikoff
(D) Joseph Lister

53. Who developed process of pasteurization during his studies on wine?
(A) Thomas Pasteur
(B) Robert Koch
(C) Louis Pasteur
(D) Richard Petri

54. Who discovered aerobic and anaerobic fermentative micro-organisms?
(A) Sergi Winogradsky
(B) Beijerinck
(C) Louis Pasteur
(D) Robert kirchoff

55. Who discovered that soil bacteria could oxidize iron, sulfur, and ammonia to obtain energy
(A) Sergei N. Winogradsky
(B) Martinus W. Beijerinck
(C) J. B. Boussingault
(D) J. V. Liebig

56. Who discovered the phenomenon of anaerobic respiration?
(A) Martinus Beijerinck
(B) Louis Pasteur
(C) Elie Metchnikoff
(D) Sergei Winogradsky

57. Who first isolated nitrogen-fixing anaerobic soil bacteria?
(A) S. N. Winogradsky
(B) M. W. Beijerinck
(C) R. Warington
(D) Robert Kirchoff

58. Who first studied decomposition of cellulose?

(A) Robert Kirchoff
(B) Charles Darvin
(C) Sergie Winogradsky
(D) Martinus Beijerinck

59. Who first isolated aerobic nitrogen fixing bacterium *Azotobacter*?

(A) Sergei N. Windogradsky
(B) Martinus W. Beijerinck
(C) Robert Kirchoff
(D) J. V. Liebig

60. Who first isolated a root nodule bacterium also capable of fixing nitrogen (later named *Rhizobium*)?

(A) S. N. Windogradsky
(B) M. W. Beijerinck
(C) R. Warington
(D) J. V. Liebig

61. Who first isolated sulphate reducing bacteria?

(A) R. Pandey
(B) S. Windogradsky
(C) M. Beijerinck
(D) L. Pasteur

62. Who developed the enrichment-culture technique and use of selective media?

(A) J. W. Gerdemann
(B) Van Niel
(C) M. Beijerinck
(D) Robert koch

63. Who is known as father of antiseptic or aseptic surgery?

(A) Joseph Meister
(B) Joseph Lister
(C) Robert Koch
(D) Louis Pasteur

64. Who developed autoclave and porcelain filters?

(A) Camberland (B) Chearland
(C) Pasteur (D) Koch

65. Who discovered process of Gram's staining?

(A) Crystal Gram
(B) Graham Gram
(C) Christian Gram
(D) Albert Gram

66. Who first filtered the virus particles from infected tobacco plants?

(A) Beijerinck (B) Ivanowsky
(C) Buchner (D) Pasteur

67. Who prepared extract of yeast?

(A) Buchner (B) Koch
(C) Pasteur (D) Windogradsky

68. Who proved that a virus particle causes the tobacco mosaic disease?

(A) Windogradsky (B) Pasteur
(C) Stanley (D) Beijerinck

69. Who discovered blood groups?

(A) Sandsteiner (B) Lechenshire
(C) Landsteiner (D) Louis Pasteur

70. Who discovered chemotherapeutic agent?

(A) Behring (B) Ehrlich
(C) Pasteur (D) Koch

71. Who discovered lysozyme?

(A) Fleming (B) Metchnikoff
(C) Waksman (D) Pasteur

72. Who discovered penicillin antibiotic?

(A) Waksman
(B) Beijerinck
(C) Fleming
(D) Windogradsky

73. Who discovered bacterial transformation?

(A) Griffith (B) Stanley
(C) Ivanovaskya (D) Ehrlich

74. Who first developed transmission electron microscope?

(A) Jacob (B) Monad
(C) Ruska (D) Aleksandrove

75. Who first crystallized tobacco mosaic virus?

(A) Ivanovaskya (B) Murray
(C) Pasteur (D) Stanley

76. Who gave one-gene-one-enzyme hypothesis?

(A) Beedle and Tatum
(B) Lederberg and Tatum
(C) Murray and Pasteur
(D) Pasteur and Koch

77. Who discovered streptomycin antibiotic?

(A) Waksman (B) Fleming
(C) Pasteur (D) Koch

78. Who first describe bacterial conjugation?

(A) Nirenberg and Khorana
(B) Lederberg and Tatum
(C) Beedle and Tatum
(D) Landsteiner

79. Who proposed the double helix structure for DNA?

(A) Davidson and Crick
(B) Watson and Grick
(C) Watson and Crick
(D) Sampson and Glick

80. Who showed that bacteriophages inject DNA into host cells - the process of transduction?

(A) Hershey and Chase
(B) Maxam and Gilbert
(C) Sanger and Nicholsan
(D) Nirenberg and Khorana

81. Who proposed the operon model of gene regulation?

(A) Jacob and Monod
(B) Beedle and Tatum
(C) Maxam and Gilbert
(D) Nirenberg and Khorana

82. Who elucidated the genetic code?

(A) Beedle and tatum
(B) Sanger and Nicholsan
(C) Robert Koch
(D) Nirenberg and Khorana

83. Who is pioneer in marine biotechnology?

(A) Mita R. Colwell
(B) Rita R. Colwell
(C) Louis Pateur
(D) Robert Koch

84. Who contributed in bacterial taxonomy and its cell envelopes?

(A) R. G. E. Murray
(B) K. Murray
(C) L. Murray
(D) A. Murray

85. Who first discovered bacterial viruses or bacteriophage?

(A) Felix d'Herelle
(B) Twort
(C) R. Colwell
(D) K Murray

86. Who established the microbiological transformation of nitrogen and sulphur?

(A) S. N. Winogradsky
(B) W. M. Stanley
(C) W. B. Leismaan
(D) M. W. Beijerinck

87. Who discovered blood grouping technique?

(A) Bandsteiner (B) Landsteiner
(C) Buchner (D) Ivanovaskya

88. Who discovered that Rocky Mountain spotted fever was caused by Rickettsia?

(A) Ricketts (B) Pasteur
(C) Koch (D) Twort

89. **Who developed the technique of DNA sequencing?**
(A) Sanger and Gilbert
(B) Beedle and Tatum
(C) None
(D) Both

90. **PCR was developed by**
(A) K. Mullis
(B) L. Mullis
(C) M. Mullis
(D) J. Mullis

91. **Which is the first vaccine developed by genetic engineering approved for human trials?**
(A) Hepatits B (B) Cowpox
(C) Small pox (D) Rabies

92. **Who showed that leguminous plants can fix atmospheric nitrogen and increase nitrogen content in the soil?**
(A) J. Von Liebig
(B) S. N. Winogradsky
(C) J. B. Boussingault
(D) W. B. Leismaan

93. **Which of the following scientist demonstrated that root nodules in legumes were formed by a specific group of bacteria?**
(A) Boussingault and Liebig
(B) Leismaan and Woronin
(C) Winogradsky and Beijerinck
(D) Hellriegel and Wilfarth

94. **The credit for isolation of nitrifying bacteria for the first time goes to...............**
(A) Boussingault (B) Beijerinck
(C) Hellriegel (D) Windogradsky

95. **Who gave first experimental evidence of elemental nitrogen fixation by microorganisms?**
(A) Jodin (B) Warington
(C) Frank (D) Windogradsky

96. **Who coined (1885) the term "Mycorrhiza" –an association of fungal symbionts with plant roots?**
(A) Jodin (B) Warington
(C) Frank (D) Windogradsky

97. **Who showed that a mutually beneficial association exists between bacteria (Rhizobia) and legume root and legumes could utilize atmospheric nitrogen?**
(A) Boussingault and Liebig
(B) Leismaan and Woronin
(C) Winogradsky and Beijerinck
(D) Hellriegel and Wilfarth

98. **Who is known as father of microbial ecology?**
(A) Hellriegel (B) Beijerinck
(C) Wilfarth (D) Windogradsky

99. **Who discovered the phenomenon of anaerobic degradation of cellulose by soil bacteria?**
(A) Jodin (B) Frank
(C) Warington (D) Omeliansky

100. **Who coined the term rhizsphere?**
(A) Hiltner (B) Ruinen
(C) Jodin (D) Frank

101. **Who developed "Direct soil examination" technique for studying soil microorganisms?**
(A) Conn (B) Beijerinck
(C) Omeliansky (D) Frank

102. **Who demonstrated importance of molybdenum in accelerating nitrogen fixation by nodulating legumes**
(A) Rossi (B) Bortels
(C) Van Niel (D) Kubo

103. **Who proved role and importance of leghaemoglobin in nitrogen fixation**
(A) Rossi (B) Bortels
(C) Van Niel (D) Kubo

104. Who reported occurrence of VAM

(A) Rossi and Cholond
(B) Barbara Mosse and Gerdemann
(C) Hardy & Associates
(D) Dobereiner and associates

105. Who developed the technique of acetylene reduction assay for estimation of biological nitrogen fixation?

(A) Rossi and Cholond
(B) Barbara Mosse and Gerdemann
(C) Hardy & Associates
(D) Dobereiner and associates

106. Who first time reported occurance of associative symbiotic *Azospirillum* bacteria?

(A) Rossi and Cholond
(B) Barbara Mosse and Gerdemann
(C) Hardy & Associates
(D) Dobereiner and associates

107. Who first reported nodulation on stem of *Sesbania rostrata*

(A) Rossi and Cholond
(B) Dommergues & associates
(C) Hardy & Associates
(D) Dobereiner and associates

108. Who developed contact slide technique

(A) Rossi and Cholond
(B) Barbara Mosse and Gerdemann
(C) Hardy & Associates
(D) Dobereiner and associates

| Answer Key |

Q. No.	Ans	Q. No.	Ans	Q. No.	Ans	Q. No.	Ans	Q. No.	Ans	Q. No.	Ans
1.	A	2.	C	3.	B	4.	B	5.	D	6.	C
7.	A	8.	B	9.	B	10.	D	11.	C	12.	A
13.	C	14.	D	15.	A	16.	B	17.	B	18.	C
19.	A	20.	C	21.	B	22.	B	23.	D	24.	A
25.	B	26.	B	27.	D	28.	A	29.	A	30.	C
31.	D	32.	C	33.	B	34.	A	35.	C	36.	B
37.	C	38.	A	39.	D	40.	D	41.	A	42.	C
43.	A	44.	A	45.	C	46.	B	47.	A	48.	C
49.	C	50.	B	51.	A	52.	A	53.	C	54.	C
55.	A	56.	D	57.	A	58.	C	59.	B	60.	B
61.	C	62.	C	63.	B	64.	A	65.	C	66.	B
67.	A	68.	D	69.	C	70.	B	71.	A	72.	C
73.	A	74.	C	75.	D	76.	A	77.	A	78.	B
79.	C	80.	A	81.	A	82.	D	83.	B	84.	A
85.	C	86.	A	87.	B	88.	A	89.	A	90.	A
91.	A	92.	C	93.	B	94.	D	95.	A	96.	C
97.	D	98.	B	99.	D	100.	A	101.	A	102.	B
103.	D	104.	B	105.	C	106.	D	107.	B	108.	A

Chapter **2**

PROKARYOTIC CELL

1. In which part of bacterial cell double-stranded single circular DNA is found?

(A) Protonucleus (B) Nucleus
(C) Nucleoid (D) Nucleoplasm

2. Prokaryotes are classified in which domains?

(A) Bacteria and Protista
(B) Bacteria and Archaea
(C) Archaea and Eukarya
(D) Eukarya and Monera

3. Which of the following organisms are most predominant in thermal vents?

(A) Eubacteria (B) Yeast
(C) Protista (D) Archaea

4. Polysaccharides are present in which of the following organelle?

(A) Cell wall (B) Pili
(C) Flagella (D) Plasmids

5. The slimy outer covering of bacteria which protects them from phagocytosis is know as

(A) Capsule (B) Cell wall
(C) Flagellum (D) Peptidoglycan

6. Which prokaryotic cell organelle contains nucleic acid?

(A) Volutine Granules
(B) Capsule
(C) Pili (D) Plasmids

7. Which of the following bacterial cell structure do not possess protein?

(A) Cell membrane (B) Flagella
(C) Plasmids (D) Pili

8. Which of the following is smallest living cell with cell wall?

(A) Bacteria (B) Viroids
(C) Mycoplasma (D) Yeast

9. Which of the following organelles is smallest in size?

(A) Ribosome
(B) Mitochondria
(C) Chloroplast
(D) Lysosome

10. Micrographia book was authored by

(A) Robert Hook
(B) E. V. Leeuwenhoek
(C) Robert Koch
(D) Louis Pasteur

11. Stickiness in bacterial cell is due to presence of

(A) Cell wall
(B) Nuclear membrane
(C) Plasma membrane
(D) Glycocalyx

12. Which of the following cell organelles is involved in energy generation from carbohydrate?

(A) Lysosome
(B) Ribosome
(C) Chloroplast
(D) Mitochondrion

13. Prokaryotic cell do not possess

(A) Nuclear envelope
(B) Ribosome
(C) Mesosome
(D) Plasma membrane

14. A protoplast is a cell
(A) Without plasma membrane
(B) Without nucleus
(C) Undergoing division
(D) Without cell wall

15. The plasma membrane consists mainly of
(A) Phospholipids embedded in a protein bilayer
(B) Proteins embedded in a phospholipid bilayer
(C) Proteins embedded in a polymer of glucose molecules
(D) Proteins embedded in a carbohydrate bilayer

16. Flagella of prokaryotic & eukaryotic cells differ in
(A) Type of movement & placement in a cell
(B) Location in cell & mode of functioning
(C) Microtubular organisation & type of movement
(D) Microtubular organisation & function

17. The jelly-like fluid substance present in cells is known as
(A) Cytoplasm (B) Protoplast
(C) Nucleosome (D) Ribosome

18. Which of the following is prokaryote?
(A) *Bacillus megaterium*
(B) *Trichoderma viride*
(C) *Trypanosoma cruzi*
(D) *Meloidogyne incognita*

19. The bacterial cells showing helical shape are known as
(A) Cocci (B) Bacillus
(C) Vibrio (D) Spirillum

20. The cells containing long multi-nucleate filaments or hyphae that may branch to produce a network called a mycelium is characteristic of
(A) Actinomycetes
(B) Spirillum
(C) Streptobacilli
(D) Streptococci

21. Flexible spiral shaped bacteria are known as
(A) Spirilla (B) Streptobacilli
(C) Actinomycetes (D) Spirochetes

22. Which among the following are the smallest bacteria having 0.3 µm diameter?
(A) Spiroplasma (B) Rikettsia
(C) Microplasma (D) Mycoplasma

23. Gram's reaction mainly differentiate bacteria based on composition of
(A) Cell membrane
(B) Cell wall
(C) Pilli
(D) Cytoplasm

24. The space between plasma membrane and cell wall of bacterial cell is known as
(A) Mesoplasmic space
(B) Plasmic space
(C) Periplasmic space
(D) Macrospace

25. Which is known as selectively permeable barrier of bacterial cell?
(A) Capsule
(B) Cell wall
(C) Plasma membrane
(D) Mesosome

26. The nutrient and waste transport in bacterial cell occurs through
(A) Cell wall
(B) Exopolysachharide
(C) Cell membrane
(D) Capsule

27. Which organelle is the site of metabolic processes of bacterial cell including respiration and photosynthesis?

(A) Plasmodesmata
(B) Cellwall
(C) Flagella
(D) Mitocondria

28. Which organelle of bacterial cell detects environmental cues for chemotaxis

(A) Flagella (B) Pilli
(C) Plasma membrane
(D) Cell wall

29. Plasma membrane contains

(A) Proteins & Carbohydrate
(B) Lipids & Proteins
(C) Proteins and Nucleic acid
(D) Nucleic acid and lipid

30. Plasma membrane lipids are........in nature

(A) Hydrophilic (B) Hydrophobic
(C) Amphipathic (D) Hydroneutral

31. Prokaryotic membrane do not contain

(A) Protein (B) Lipid
(C) Sterols (D) Alkanes

32. Sterol like compound present in bacterial plasma membrane are known as

(A) Hopanoids (B) Propanoid
(C) Hexanoids (D) Sterenoids

33. Function of hopanoids in bacterial membrane

(A) Nutrient transport
(B) Metabolism
(C) Respiration
(D) Stabilization of membrane

34. Thickness of cell membrane is

(A) 5-10 µm (B) 5-10 nm
(C) 5-10 cm (D) 5-10 mm

35. Ultra structure of bacterial cell membrane can be seen with

(A) Phase contrast microscope
(B) Electron microscope
(C) Dark field microscope
(D) Confocal microscope

36. The most widely accepted model of plasma membrane structure is

(A) Singer mosaic model
(B) Fluid mosaic model
(C) Flood mosaic model
(D) Sanger Nicholsan model

37. Most widely accepted model for cell membrane structure was proposed by

(A) S. Jonathan Singer and Garth Nicholson
(B) Louis Pasteur
(C) Francesco Redi
(D) Theodore Schwann

38. Proportion of peripheral protein in plasma membrane is

(A) 30-40 % (B) 50-60 %
(C) 20-30 % (D) 70-80 %

39. Amount of integral protein in plasma membrane is

(A) 20-30 % (B) 50-60 %
(C) 70-80 % (D) 80-100 %

40. Invagination of plasma membrane in the form of vesicles, tubules and lamella is known as

(A) Glycocalyx (B) Mesosomes
(C) Mitocondria (D) Chlorophyll

41. Which organelle plays an important role in cell wall formation?

(A) Septosome (B) Chromosome
(C) Mesosomes (D) Membrane

42. Which organelle plays an important role in chromosome replication and distribution to daughter cells

(A) Chromosome (B) Cell wall
(C) Cytoplasm (D) Mesosomes

43. **The substance lying between plasma membrane and nucleoid is known as**
(A) Chromosome
(B) Ribosome
(C) Cytoplasmic matrix
(D) Cytoplasmic membrane

44. **Bacterial cell contains ....% water**
(A) 70 (B) 80
(C) 90 (D) 99

45. **In bacterial cell all the cellular material bounded by plasma membrane is known as**
(A) Spheroplast
(B) Protoplast
(C) Cytoplasm
(D) Matrix

46. **Granules of organic and inorganic material present in the cytoplasmic matrix of bacterial cell is known as**
(A) Storage bodies
(B) Protoplasmic bodies
(C) Inclusion bodies
(D) Cytoplasmic bodies

47. **The organelle meant for food storage in bacterial cell are**
(A) Inclusion bodies
(B) Cytoplasm
(C) Periplasm
(D) Mitochondria

48. **Which of the following inclusion body is not membrane bound?**
(A) Polyphosphate, Cyanophycin and Glycogen granules
(B) PHB, Glycogen and Sulphate granules
(C) Carboxysome, Magnetosme and PHB
(D) Carboxysome, Phosphate and Sulphate granules

49. **Membrane bound inclusion bodies are**
(A) poly-hydroxybutyrate granules (PHB)
(B) Polyphosphate granules
(C) Cyanophycin granules
(D) Glycogen granules

50. **Which of the following are organic inclusion bodies?**
(A) Glycogen, PHB and Cyanophycin granules
(B) Carboxysome, Mangnetosome and PHB granules
(C) Carboxysome, PHB and Cyanophysin
(D) Phopshate, Sulphate and Glycogen granules

51. **In which bacteria both glycogen and PHB granules are present?**
(A) Green sulfur bacteria
(B) Cyanobacteria
(C) Purple photosynthetic bacteria
(D) Green bacteria

52. **PHB granules can be easily stained with which dye?**
(A) Crystal violet (B) India ink
(C) Sudan black (D) Safranin

53. **Carbon storage granules in bacterial cells are**
(A) Glycogen and PHB
(B) Cyanophycin and Glycogen
(C) Polyphosphate and PHB
(D) Cyanophycin and Polyphosphate

54. **Inclusion bodies which contain large polypeptide are**
(A) PHB granules
(B) Carboxysome
(C) Cyanophycin granules
(D) Proteionic granules

55. Prime nitrogen storage granule in bacteria are
(A) Cyanophycin granules
(B) PHB granule
(C) Carboxysome
(D) Gas vesicle

56. Carboxysomes are found in
(A) Cyanobacteria, Nitrifying bacteria and Thiobacillus
(B) Nitrifying, Denitrifying and Nitrogen fixing bacteria
(C) Thiobacilli, Sulfolobus and Cyanobacteria
(D) Iron, Sulphur and Denitrifying bacteria

57. Carboxysomes serves as storehouse of enzyme
(A) Ribose -1,5- bis phosphate carboxylase
(B) Ribulose-1,5-bisphosphate carboxylase
(C) Ribose -1,5-bis phosphate decarboxylase
(D) Ribulose-1,5-bis phosphate decarboxylase

58. Site of CO_2 fixation in bacteria is
(A) PHB granules
(B) Carboxysome
(C) Ribosome
(D) Mesosome

59. Gas vacuoles are present in which of the following groups of bacteria?
(A) Cyanobacteria, Photosynthetic bacteria and Thiothrix
(B) Iron, Sulphur and Denitrifying bacteria
(C) Thiobacillus, Thiothrix and Pseudomonas denitrificans
(D) Cyanobacteria, Thiobacillus and Iron bacteria

60. Buoyancy of bacterial cell is due to presence of
(A) Cyanophycin granules
(B) Gas vacuoles
(C) PHB granule
(D) Carboxysome

61. Gas vacuoles contains
(A) Proteins (B) Lipid
(C) Carbohydrate (D) Nucleic acid

62. Name inorganic inclusion bodies in bacterial cells
(A) Carboxysomes, Magnetosome and PHB granules
(B) Cyanophycin, Sulphate and PHB granules
(C) Magnetosomes, Polyphosphate and Sulphate granules
(D) Cyanophycin, Sulphate and PHB

63. Sugars present in peptidoglycan are
(A) N-acetylglycosamine and N-acetylnuramic acid
(B) N-acetylglucosamine and N-acetylmuramic acid
(C) N-acetylglycosamine and N-acetylmuramic acid
(D) N-acetylgalactosamine and N-acetylnuramic acid

64. Polyphosphate granules are also known as
(A) Volutine or Metachromatic granules
(B) Volatile or Metachromatic granules
(C) Volutine or Monochromatic granules
(D) Volatile or Monochromatic granules

65. Volutine granules are storage of
(A) Carbon (B) Sulfur
(C) Phosphate (D) Hydrogen

66. Sulphur granules are present in
(A) Cyanobacteria
(B) Green photosynthetic bacteria
(C) Purple photosynthetic bacteria
(D) Iron oxidizing bacteria

67. Which inclusion bodies are used by some bacteria to orient in the earth's magnetic field?
(A) Magnetosome
(B) Carboxysomes
(C) Polyphosphate granules
(D) Sulphate granules

68. Ribosomes can be seen in
(A) Light microscope
(B) Electrone microscope
(C) Dark field microscope
(D) Bright field microscope

69. Ribosomes are generally made up of
(A) Proteins and RNA
(B) RNA and DNA
(C) Protein and Lipid
(D) Protein and DNA

70. Site of protein synthesis in prokaryotic cell is
(A) Mitochondria
(B) Ribosomes
(C) Mesosomes
(D) Carboxysome

71. Which proteins play an important role in proper folding of newly synthesized Proteins
(A) Chaperones
(B) Heat shock proteins
(C) Histone
(D) Hapanoids

72. Bacterial cell contain which type of ribosome?
(A) 60 S (B) 70 S
(C) 80 S (D) 50 S

73. Which prokaryote (bacteria) contains more than one chromosome
(A) *Bacillus subtilis*
(B) *Vibrio cholerae*
(C) *Salmonella typhi*
(D) *Azospirillum lipoferum*

74. Nucleoids are visible in
(A) Light microscope
(B) Electron microscope
(C) Dark field microscope
(D) Phase contrast microscope

75. Extra chromosomal DNA in bacterial cell is known as
(A) Plastid (B) Plasmid
(C) Chromosome (D) Nucleoid

76. Self (autonomous) replicating piece of DNA of bacterial cell is known as
(A) Plasmid
(B) Chromosome
(C) Nucleoid
(D) Plasmodesmata

77. The genes which give bacterial cell a selective advantage are present in
(A) Plasmid
(B) Chromosome
(C) Chloroplast
(D) Mitocodria

78. The organelle which give shape to the bacterial cell and protects it from osmotic shock is
(A) Cell membrane
(B) Cell wall
(C) Periplasm
(D) Periplasmic space

79. Cell wall is mainly composed of
(A) Peptidoglycan or murein
(B) Lipid
(C) Carbohydrate
(D) Protein

80. Who developed procedure of Gram's staining
(A) Thomas Gram 1985
(B) Christian Gram 1884
(C) Christian Gram 1985
(D) Thomas Gram 1884

81. Size of peptidoglycan layer in gram +ve bacteria is
(A) 2-7 nm (B) 20-80 nm
(C) 10-50 nm (D) 20-100 nm

82. Periplasmic space contains which type of proteins?
(A) Transport (B) Spacer
(C) Cytoplasmic (D) Histone

83. Peptidoglycan is
(A) Lipid (B) Protein
(C) Polymer (D) Nucleotide

84. Peptidoglycan contains
(A) N-acetylglucosamine, N-acetylmuramic acid and D amino acids
(B) N-acetylglycosamine, L-acetylmuramic acid and L amino acids
(C) N-acetylglucoṣamine, N-acetylmuramic acid and L amino acids
(D) N-acetylglycosamine, D-acetylmuramic acid and D amino acids

85. N-acetylglucosamine and N-acetylmuramic acid are attached with each other by
(A) Hydrogen bond
(B) Peptide interbridge
(C) Carboxylic bond
(D) Glycosidic bond

86. Gram -ve cell walls do not contain
(A) Peptidoglycan (B) Teichoic acids
(C) Lipid (D) Sterols

87. Gram +ve cells do not contain
(A) Teichoic acid
(B) Outer membrane
(C) Peptidoglycan
(D) Proteins

88. Gram –ve cell's outer membrane contains
(A) Porous protein
(B) Sterol
(C) Braun's lipoproteins
(D) Teichoic acid

89. Outer membrane is more permeable to small molecules such as glucose and other monosachharides due to presence of
(A) Loose bonding of lipid bilayer
(B) Porin proteins
(C) Less peptidoglycan content
(D) All of above

90. The process in which water moves from dilute solutions to more concentrated solutions is known as
(A) Osmolysis
(B) Osmosis
(C) Reverse osmosis
(D) Osmoplysis

91. If a bacterial cell is placed in hypotonic solution than water moves
(A) Inside the cell
(B) Outside the cell
(C) Solute moves inside the cell
(D) Solute moves outside the cell

92. In hypertonic solution water moves
(A) Inside the cell
(B) First outside and then inside cell
(C) Outside the cell
(D) First inside and then outside cell

93. Spheroplast are produced from G –ve cell by treatment with
(A) Penicillin (B) Streptomycin
(C) Tetracyclin (D) Erythromuycin

94. **Cell wall less bacteria are**
(A) Spiroplasma (B) Mycoplasma
(C) Microplasma (D) Spheroplasma

95. **Mycoplasma occurs in .... shape**
(A) Rod (B) Round
(C) Spiral (D) Pleomorphic

96. **Capsule contains**
(A) Lipid (B) Carbohydrate
(C) Polysachharide (D) Protein

97. **Which organelle protect bacteria against phagocytosis and desiccation**
(A) Cell wall (B) Capsule
(C) Flagella (D) Pilli

98. **Short, fine hair like appendages that are not involved in motility are known as**
(A) Flagella
(B) Pilli
(C) Axial filament
(D) Fimbriae

99. **Fimbriae are visualized by**
(A) Simple microscope
(B) Dark field microscope
(C) Bright field microscope
(D) Electron microscope

100. **The main function of fimbriae is**
(A) Attachment (B) Locomotion
(C) Conjugation (D) Gliding

101. **Sexual appendages in bacteria are**
(A) Flagella (B) Pilli
(C) Fimbriae (D) Capsule

102. **How many pilli are present per cell**
(A) 1000 (B) 2-20
(C) 400 (D) 1-10

103. **Which appendages plays an imp role in bacterial mating or conjugation**
(A) Pilli (B) Flagella
(C) Capsule (D) Fimbriae

104. **In G -ve bacterial flagella how many basal body rings are present**
(A) One (B) Two
(C) Three (D) Four

105. **In gram +ve bacteria how many basal body rings are present**
(A) Two (B) Three
(C) Four (D) Five

106. **The forward movement of bacteria is known as**
(A) Run
(B) Tumbling
(C) Gliding
(D) Swimming

107. **When flagella move anticlockwise than bacterial cell moves..............**
(A) Forward (B) Reverse
(C) Stop (D) Glide

108. **The clockwise movement of bacterial flagella results in**
(A) Running
(B) Tumbling
(C) Stopping
(D) Backwarding

109. **Spirochetes (spiral bacteria) move by means of**
(A) Flagella
(B) Pilli
(C) Axial filament
(D) Capsule

110. **Axial filament is composed of**
(A) Periplasmic flagella
(B) Pilli
(C) Cytoplasmic flagella
(D) Screw flagella

111. **Gliding motility is seen in**
(A) Cyanobacteria
(B) Myxobacteria
(C) Cytophaga
(D) All of above

112. Which are the locomotory appendages in bacteria

(A) Flagella and axial filament
(B) Flagella and cilia
(C) Flagella and pili
(D) Pili and cilia

113. Resting structures found in G +ve bacteria are

(A) Cyst (B) Glycocalyx
(C) Spore (D) Carboxysomes

114. Resting structure found in G -ve bacteria are

(A) Spore (B) Cyst
(C) Capsule (D) Slime layer

115. Which organelles are impermeable to stains

(A) Cell wall (B) Cell membrane
(C) Spores (D) Capsule

116. Heat resistant structures of bacterial cells are

(A) Cell wall (B) Spores
(C) Capsule (D) Cyst

117. Heat resistance of spores is due to presence of which of the following compound?

(A) Glutamic acid
(B) N acetyl glucosamine
(C) Dipicolinic acid
(D) N acetyl muramic acid

118. Which of the following is likely to contain structures made up of N-acetylmuramic acid and N-acetylglucosamine

(A) Spheroplasts (B) Mycoplasmas
(C) Escherichia coli (D) Protoplasts

119. High resistance of bacterial cells to osmotic shock as compared to eukaryotes is due to

(A) Cellulose present in cell wall
(B) Presence of osmo- regulating porins
(C) Selectively permeable nature
(D) Peptidoglycan in cell wall

120. The use of the antibiotics affecting ribosome will result in the loss of which of the following functions?

(A) ATP production
(B) DNA replication
(C) Phagocytosis.
(D) Protein synthesis

121. Which of the following organelle serves as mechanical boundary of cell?

(A) Capsule
(B) Glycocalyx
(C) Exospore
(D) Plasma membrane

| Answer key |

Q. No.	Ans	Q. No.	Ans	Q. No.	Ans	Q. No.	Ans	Q. No.	Ans	Q. No.	Ans
1.	C	2.	B	3.	D	4.	A	5.	A	6.	D
7.	C	8.	A	9.	A	10.	A	11.	D	12.	D
13.	A	14.	D	15.	B	16.	C	17.	A	18.	A
19.	D	20.	A	21.	D	22.	D	23.	B	24.	C
25.	C	26.	C	27.	A	28.	C	29.	B	30.	C
31.	C	32.	A	33.	D	34.	B	35.	B	36.	B
37.	A	38.	C	39.	C	40.	B	41.	C	42.	D
43.	C	44.	A	45.	B	46.	C	47.	A	48.	A
49.	A	50.	A	51.	C	52.	C	53.	C	54.	C
55.	A	56.	A	57.	B	58.	B	59.	A	60.	B
61.	A	62.	C	63.	B	64.	A	65.	C	66.	C
67.	A	68.	B	69.	A	70.	B	71.	A	72.	B
73.	B	74.	B	75.	B	76.	A	77.	A	78.	B
79.	A	80.	B	81.	B	82.	A	83.	C	84.	A
85.	B	86.	B	87.	B	88.	C	89.	B	90.	B
91.	A	92.	C	93.	A	94.	B	95.	D	96.	C
97.	B	98.	D	99.	D	100.	A	101.	B	102.	D
103.	A	104.	D	105.	A	106.	A	107.	A	108.	B
109.	C	110.	A	111.	D	112.	A	113.	C	114.	B
115.	C	116.	B	117.	C	118.	C	119.	D	120.	D
121.	D										

◈

Chapter 3

EUKARYOTIC CELL

1. **Which of the following is eukaryote?**
(A) Fungi (B) Bacteria
(C) Virus (D) Archea

2. **Which organelles are not membrane bound in eukaryotic cells?**
(A) Mitochondria (B) Ribosomes
(C) Plastid (D) Golgibody

3. **In eukaryotic cells which organelle are involved in cell motion and shape change?**
(A) Flagella (B) Axial filament
(C) Microfilament (D) Pilli

4. **Microtubules helps in**
(A) Maintainance of cell shape
(B) Help in movement
(C) Intracellular transport
(D) All of above

5. **Granular endoplasmic reticulum contains**
(A) Microtubules
(B) Microfilaments
(C) Ribosomes
(D) Starch granules

6. **Role of endoplasmic reticulum**
(A) Lipid and protein synthesis
(B) Sugar and protein synthesis
(C) Lipid and sugar synthesis
(D) Polysaccharide and polypeptide synthesis

7. **Flattened stacks of cisternae present in Golgi apparatus are known as**
(A) Dictiosome (B) Golgisome
(C) Carboxysome (D) Mesosomes

8. **Functions of golgi apparatus**
(A) Packaging and secretion of materials
(B) Lysosome formation
(C) Mesososme formation
(D) Carboxysome formation

9. **The organelle involved for intra-cellular digestion in eukaryotes are**
(A) Ribosome (B) Carboxysome
(C) Lysosomes (D) Mitochondria

10. **Lysosomes contain which type of enzymes**
(A) Lipase (B) Hydrolases
(C) Protease (D) Oxidoreductase

11. **Enzymes of lysosomes work best at which pH**
(A) Acidic (B) Basic
(C) Neutral (D) Alkaline

12. **The process in which cells take up solutes or particles by enclosing them vacuoles of vesicles pinched of from its plasma membrane**
(A) Osmosis (B) Endocytosis
(C) Exocytosis (D) Diffusion

13. **Which of the following is involved in endocytosis**
(A) Endosomes, Phagosome and Pinosome
(B) Mesosome, Carboxysome and Phagosome
(C) Carboxysome, Pinosome and Endosome
(D) Phagosome, Mesosome, Carboxysome

14. **The process in which cells selectively digest its own portion in known as**
(A) Endophagy (B) Selfphagy
(C) Autophagy (D) Exophagy

15. **Recycling of cell constituents occurs through**
(A) Autophagy (B) Cytophagy
(C) Phagocytosis (D) Endophagy

16. **Digestive vacuoles in eukaryotic cells are**
(A) Lysosome (B) Phagosome
(C) Pinosome (D) Endosome

17. **Non-lysosomal protein digestive system found in eukaryotic cells is known as**
(A) Phagosome (B) Pinosome
(C) Proteososmes (D) Carboxysome

18. **Eukaryotic ribosomes are**
(A) 80S (B) 70S
(C) 40S (D) 60S

19. **Eukaryotic ribosomes are dimer of**
(A) 40 S and 60S (B) 30S and 70S
(C) 40S and 40S (D) 30S and 50S

20. **Protein synthetic machinery of cell is**
(A) Mitochondria
(B) Ribososme
(C) Carboxysome
(D) Endoplasmic reticulum

21. **Powerhouse of cell where ATP synthesis occurs**
(A) Endoplasmic reticulum
(B) Ribosome
(C) Mitochondria
(D) Plastid

22. **Folding of inner mitochondrial membranes are known as**
(A) Cisternae (B) Cristae
(C) Microtubules (D) Microfilament

23. **Which eukaryotic organelle contains 70S ribosome**
(A) Mitochondria and Chloroplast
(B) Mitochondria and Chromosome
(C) Mitochondria and Carboxysome
(D) Chloroplast and Carboxysome

24. **Which of the following eukaryotic organelle resembling prokaryotic cell multiply by binary fission**
(A) Mitochondria and Chloroplast
(B) Mitochondria and Chromosome
(C) Mitochondria and Carboxysome
(D) Chloroplast and Carboxysome

25. **Cytoplasmic organelles of algae and higher plants that often possess pigments such as chlorophylls and carotenoids, and are the sites of synthesis and storage of food reserves is known as**
(A) Chlorosome (B) Chromosome
(C) Plasmid (D) Plastid

26. **Chloroplast is an important type of**
(A) Pigment (B) Membrane
(C) Plastid (D) Plasmid

27. **The matrix (core) of the chloroplast is known as**
(A) Grana (B) Stroma
(C) Plastid (D) Membrane

28. **Membrane delimited sac like structures of chloroplast are known as**
(A) Grana (B) Stroma
(C) Plastid (D) Thylakoids

29. **Several disk like thylakoids are stacked on each other like coins forming structures known as**
(A) Plastid (B) Grana
(C) Stroma (D) Chloroplast

30. **The process of formation of carbohydrate from CO_2 and water is known as**
(A) Dark reaction
(B) Light reaction
(C) Photocarbo synthesis
(D) Phosphorylation

31. Dark reaction during photosynthesis takes place in which potion of chloroplast

(A) Grana

(B) Thylakoid

(C) Stroma

(D) Membrane

32. The process of trapping of light energy to generate ATP, NADPH, and O_2 is known as

(A) Dark reaction

(B) Calvin reaction

(C) Photophosphorylation

(D) Light reaction

33. Light reaction of photosynthesis takes place in

(A) Grana

(B) Stroma

(C) Thylakoid membrane

(D) Plastid

34. Many algal chloroplast contains dense region of protein surrounded by starch or another Polysaccharide which is known as

(A) Thylakoid (B) Grana

(C) Stroma (D) Pyrenoids

35. Function of pyrenoids is

(A) Protein synthesis

(B) Lipid ssynthesis

(C) Polysaccharide synthesis

(D) ATP synthesis

36. The repository (storehouse) of eukaryotic cell's genetic information and its control centre is

(A) Plasmid (B) Chloroplast

(C) Nucleolus (D) Nucleus

37. The inner matrix of nucleus is known as

(A) Chromoplasm (B) Nucleoplasm

(C) Chromosome (D) Cytoplasm

38. The DNA containing part of nucleus is known as

(A) Nucleoid (B) Nucleosome

(C) Nucleolus (D) Chromatin

39. Loosely organized chromatin are known as

(A) Euchromatin

(B) Homochromatin

(C) Heterochromatin

(D) Chromatin

40. The genes that express themselves actively are present on

(A) Homochromatin

(B) Euchromatin

(C) Heterochromatin

(D) Chromomere

41. Tightly coiled chromatin are known as

(A) Heterochromatin

(B) Homochromatin

(C) Euchromatin

(D) Chromomere

42. Which portion of the nucleus is not genetically active all the time

(A) Euchromatin

(B) Heterochromatin

(C) Homochromatin

(D) Plasmid

43. The prime function of nucleolus in eukaryotes is

(A) DNA replication

(B) Ribosome synthesis

(C) Protein synthesis

(D) Synthesis of ribosomal RNA (rRNA)

44. The process of cell division in which number of chromosomes remain similar after division is known as

(A) Mitosis (B) Meiosis

(C) Binary fission (D) None

45. The process by which the number of chromosomes is reduced to half with each daughter cell receiving one complete set of chromosomes is called as

(A) Mitosis (B) Meiosis
(C) Duplication (D) Reduction

46. Growth cycle of eukaryotic cell contains how many phases

(A) Two (B) Three
(C) Four (D) Five

47. In which phase of growth cycle synthesis of mRNA, tRNA, ribosomes and cytoplasmic constituents occurs and nucleolus grows rapidly

(A) G1 phase (B) S phase
(C) G2 phase (D) M phase

48. In which phase of growth cycle rapid synthesis and doubling of nuclear DNA and histones occurs

(A) G1 phase (B) S phase
(C) G2 phase (D) M phase

49. In eukaryotes motility is governed by

(A) Cilia and Flagella
(B) Flagella and Axial filament
(C) Axial filament and Pili
(D) Flagella and Fibriae

50. How many types of eukaryotic flagella are there

(A) One (B) Two
(C) Three (D) Four

51. The flagella containing lateral hairs called flimmer filaments are known as

(A) Tinsel
(B) Whiplash
(C) Hairy
(D) Fimbriated

52. The flagella which do not contain flimmer filament are known as

(A) Tinsel (B) Whiplash
(C) Naked (D) Hairless

53. Arrangement of microtubules 9+2 type is seen in

(A) Eukaryotic cilia
(B) Eukaryotic nucleus
(C) Eukaryotic flagella
(D) Eukaryotic chromosome

54. Eukaryotic cell wall lacks

(A) Lipid (B) Peptidoglycan
(C) Ribosome (D) RNA

55. Which of the following organelle do not contain ribosome?

(A) Mitocondria (B) Nucleus
(C) Endoplasmic reticulum
(D) Chloroplast

56. Polysomes are

(A) Multiple units of ribosome
(B) Attachment of multiple units of ribosome to a common mRNA
(C) Attachment of multiple mRNA to a ribosome
(D) Lysosomal aggregation

57. Which ion plays an important role in binding of ribosomal subunits?

(A) Na^+ (B) Mg^{+2}
(C) Mn^{+2} (D) Fe^{+2}

58. Microtubules are associated with which part of eukaryotic cell?

(A) Centrosome (B) Lysosome
(C) Ribosome (D) Peroxisome

59. DNA processing in eukaryotic cells occurs in

(A) Cytoplasm
(B) Golgi apparatus
(C) Mitochondria
(D) Vacuoles

60. Cellular respiration in eukaryotic cell occurs in

(A) Golgi apparatus
(B) Mitochondria
(C) Cytoplasm
(D) Vacuoles

61. **Glycoproteins and glycolipids are synthesized in**
(A) Lysosome
(B) Vacuoles
(C) Plastid
(D) Golgi apparatus

62. **Which one of the following is functional unit of golgi apparatus?**
(A) Thylakoid
(B) Archoplasm
(C) Cisternae
(D) Cristae

63. **Longest cells in human body are**
(A) Leg muscle cells
(B) Bone cells
(C) Nerve cells
(D) Heart muscle cells

64. **Which of the following organelle is not bounded by membrane?**
(A) Endoplasmic reticulum
(B) Lysosome
(C) Ribosomes
(D) Peroxisome

65. **Plasma membrane is composed of**
(A) Glycoproteins
(B) Lipoproteins
(C) Chromoproteins
(D) Lipids

66. **Which of the following organelle has a continuous connection with nuclear membrane?**
(A) Golgi apparatus
(B) Lysosome
(C) RER (D) SER

67. **Cristae in mitochondria function as sites for**
(A) Protein synthesis
(B) Phosphorylation of flavoproteins
(C) Breakdown of macromolecules
(D) Oxidation–reduction reaction

68. **Which of the following phospholipid is abundant in mitochondrial inner membrane?**
(A) Phosphatidyl inositol
(B) Phosphatidyl serine
(C) Cardiolipin
(D) Phosphatidyl choline

69. **Which of the following is NOT a function of mitochondrion**
(A) Electron transport and associated ATP production
(B) Fatty acid breakdown
(C) Non-shivering thermogenesis
(D) Glycolysis and associated ATP production

70. **The term mitochondria was coined by**
(A) Kolliker (B) Benda
(C) Fleenming (D) Robert Brown

71. **The number of nuclear pores is dependent on**
(A) Size of cells
(B) Transcriptional activity of the cell
(C) DNA content of the cell
(D) Replication of DNA

72. **In chromatin DNA Protein ratio is**
(A) 3:1 (B) 2:1
(C) 1:1 (D) 4:1

73. **Which of the following is function of nucleolus**
(A) RNA synthesis
(B) DNA synthesis
(C) Histone synthesis
(D) Ribosomal subunit synthesis

74. **Which of the following do not contain ribosomes**
(A) Nucleus
(B) mitochondrion
(C) chloroplast
(D) cytoplasm

75. In 70S ribosome 'S' stands for
(A) S.I unit
(B) Solubility factor
(C) Svedberg unit
(D) Solubility unit

76. Ribosomes contains
(A) RNA only
(B) RNA and Proteins
(C) RNA,DNA and Proteins
(D) Nucleic acids, proteins and lipids

77. Rough endoplasmic reticulum is well developed in cells actively engaged in
(A) Protein synthesis
(B) Nucleotide synthesis
(C) Lipid synthesis
(D) Secretory functions

78. Microtubules, motor proteins and actin filaments are part of
(A) The mechanism of photosynthesis that occurs in chloroplasts
(B) The rough ER (endoplasmic reticulum) in prokaryotic cells
(C) The cytoskeleton of eukaryotic cells
(D) The process that moves small molecules across cell membranes

79. Microfilaments are made of
(A) RNA (B) Protein
(C) DNA (D) Enzyme

80. Which of the following structure connects two plant cells?
(A) Cell wall
(B) Plasma membrane
(C) Plasmodesmata
(D) Fibrils

81. The hydrophobic ends of phospholipid molecules are:
(A) Polar (B) Non-polar
(C) Neutral (D) Bipolar

82. Which of the following eukaryotic structures are having membrane
(A) Chromosomes, nuclei and mitochondria
(B) Cytoplasm, chloroplasts and mitochondria
(C) Cytoplasm, nuclei and starch grains
(D) Chromosomes, chloroplasts and starch grains

83. Mitochondrion is bounded by
(A) A single unit membrane
(B) No membranes
(C) Plasma membranes
(D) Two unit membranes

84. Synthesis of new mitochondria occurs
(A) De novo
(B) From plasma membrane
(C) From nuclear envelop
(D) By replication

85. In which part of the mitochondria ATPase enzyme is located?
(A) Oxysomes
(B) Outer membrane
(C) Inner membranes
(D) Matrix

86. Mitochondrial DNA is
(A) Circular, single molecule and double stranded
(B) Linear, single molecule and double stranded
(C) Circular, single molecule and single stranded
(D) Linear, single molecule and single stranded

87. Metabolic enzymes for synthesis of chemical components of unit membrane are found in
(A) Rough Endoplasmic reticulum
(B) Smooth Endoplasmic reticulum
(C) Lysosomes
(D) Mitochondria

88. Which of the following forms the nuclear envelope during telophase?
(A) Cytoskeleton
(B) Centriole
(C) Golgi complex
(D) Endoplasmic reticulum

89. Which organelle serves as opening for pores of cell membrane and outer membrane of nuclear envelope?
(A) Golgi apparatus
(B) Mitochondria
(C) Endoplasmic reticulum
(D) Lysosome

90. Dictyosome is also known as
(A) Lysosome (B) Mitochondria
(C) Golgi body (D) Ribosome

91. Which of the following carry out cell secretion?
(A) Nucleolus (B) Plastids
(C) Endoplasmic reticulum
(D) Golgi complex

92. In which of the following organelles protein modification occurs?
(A) Endoplasmic reticulum and Lysosome
(B) Golgi complex and Mitochondria
(C) Endoplasmic reticulum and Golgi complex
(D) Mitochondria and Lysosome

93. Lysosomes arise from
(A) Smooth Endoplasmic Reticulum and Golgi complex
(B) Rough Endoplasmic Reticulum and Golgi complex
(C) Mitochondria and Golgi complex
(D) Lysosome and Golgi complex

94. Autophagic vesicles digest
(A) Pinosome contents
(B) Cell organelles
(C) Phagosome contents
(D) Microorganisms

95. The pattern of organization in centriole is
(A) $9 + 0$ (B) $9 + 1$
(C) $9 + 2$ (D) $9 + 3$

96. Centriole occurs as
(A) Singly (B) Pairs
(C) Threes (D) Fours

97. Function of centriole is related with
(A) Initiation of cell division
(B) Formation of cell plate
(C) Formation of nucleolus
(D) Formation of spindle fibers

98. Which of the following is forming microtubules in cilia and flagella ?
(A) Actin (B) Myosin
(C) Elastin (D) Tubulin

99. Nuclear membrane is
(A) Double, non-porous
(B) Single, non-porous
(C) Double, porous
(D) Single, porous

100. Which of the following is predominant in Nucleolus?
(A) DNA and proteins
(B) DNA and lipids
(C) RNA and proteins
(D) RNA and lipids

101. Which of the following function is facilitated by nuclear membrane?
(A) Nucleocytoplasmic exchange of materials
(B) Synapses of homologous chromosomes
(C) Anaphasic separation of daughter chromosomes
(D) Organization of spindles

102. Nucleoplasm is continuous with cytoplasm through
(A) Centriole (B) Nucleopores
(C) Endoplasmic Recticulum
(D) Golgi Body

103. The major component of the nucleus is

(A) DNA (B) RNA
(C) Lipids (D) Proteins

104. Nucleolar organizer is associated with

(A) Synthesis of plasma membrane
(B) Ribosome formation
(C) G6PD
(D) Disappearance of nuclear membrane

105. Chromosomes are best seen in which phase of cell cycle?

(A) Interphase
(B) Metaphase
(C) Prophase
(D) Telophase

106. Chromosomes with equal arms are known as

(A) Submetacentric
(B) Metacentric
(C) Telocentric
(D) Acrocentric

107. Four histones complexed with DNA forming an octamer is known as

(A) Nucleosome
(B) Centrosome
(C) Chromosome
(D) Endosome

108. Which one of the following represents proper sequence of cell cycle?

(A) S, M, G1, G2
(B) M, G1, G2, S
(C) G1, S, G2, M
(D) S, G1, G2, M

109. Nuclear membrane disappears at which stage of cell cycle?

(A) Metaphase
(B) Early prophase
(C) Anaphase
(D) Late prophase

110. In cell cycle DNA replication takes place in which phase?

(A) M-phase (B) S-phase
(C) G1 -phase (D) G2-phase

| Answer Key |

Q. No.	Ans	Q. No.	Ans	Q. No.	Ans	Q. No.	Ans	Q. No.	Ans	Q. No.	Ans
1.	A	**2.**	B	**3.**	C	**4.**	D	**5.**	C	**6.**	A
7.	A	**8.**	B	**9.**	C	**10.**	B	**11.**	A	**12.**	B
13.	A	**14.**	C	**15.**	A	**16.**	A	**17.**	C	**18.**	A
19.	A	**20.**	B	**21.**	C	**22.**	B	**23.**	A	**24.**	A
25.	D	**26.**	C	**27.**	B	**28.**	D	**29.**	B	**30.**	A
31.	C	**32.**	D	**33.**	C	**34.**	D	**35.**	C	**36.**	D
37.	B	**38.**	D	**39.**	A	**40.**	B	**41.**	A	**42.**	B
43.	D	**44.**	A	**45.**	B	**46.**	C	**47.**	A	**48.**	B
49.	A	**50.**	B	**51.**	A	**52.**	B	**53.**	C	**54.**	B
55.	B	**56.**	B	**57.**	B	**58.**	A	**59.**	A	**60.**	B
61.	D	**62.**	C	**63.**	C	**64.**	C	**65.**	B	**66.**	C
67.	D	**68.**	C	**69.**	D	**70.**	B	**71.**	B	**72.**	C
73.	D	**74.**	A	**75.**	C	**76.**	B	**77.**	A	**78.**	C
79.	B	**80.**	C	**81.**	D	**82.**	B	**83.**	D	**84.**	D
85.	A	**86.**	A	**87.**	B	**88.**	D	**89.**	C	**90.**	C
91.	D	**92.**	C	**93.**	A	**94.**	B	**95.**	A	**96.**	B
97.	D	**98.**	D	**99.**	C	**100.**	C	**101.**	A	**102.**	B
103.	D	**104.**	B	**105.**	B	**106.**	B	**107.**	A	**108.**	C
109.	B	**110.**	B								

◈

Chapter **4**

ARCHAEBACTERIA

1. **The microorganisms which are found in extreme environments are**
   (A) Archaebacteria
   (B) Hyper prokaryotes
   (C) Extreme prokaryotes
   (D) Protista

2. **Gram negative archaea lack**
   (A) Outer membrane and Complex pepdiglycan
   (B) Cell wall and Complex pepdiglycan
   (C) Cell membrane and Cell wall
   (D) Outer membrane and Cell wall

3. **Surface layer of Gram -ve archaea contains**
   (A) Carbohydrate and Glycoprotein
   (B) Lipid bilayer and Glycoprotein
   (C) Protein and Glycoprotein subunits
   (D) Protein and Polysaccharides

4. **Components of bacterial cell wall which are not present in archaeal cell wall are**
   (A) L aminoacid and Muramic acid
   (B) D-amino acid and Muramic acid
   (C) Lipid and Muramic acid
   (D) Glycoproteins and Muramic acid

5. **Instead of murein layer of bacterial cell wall archaeal cell wall contains**
   (A) Pseudomurein
   (B) Pseudoprotein
   (C) Pseudopolypeptide
   (D) Pseudopectin

6. **Archaeal cell wall contain**
   (A) N-aceetyl muramic acid
   (B) N-acetyl glucosamine
   (C) D- aminoacid
   (D) *N*-acetyltalosaminuronic acid

7. **Amino acids present in cell wall of archaea are**
   (A) D- amino acid
   (B) L- amino acids
   (C) Both in equal proportion
   (D) 20% D amino acid and 80% L amino acid

8. **Which type of bonds are present in archaeabacterial cell wall**
   (A) β 1-3 glycosidic bond
   (B) β 1-4 glycosidic bond
   (C) β 1-3 glucosidic bond
   (D) β 1-4 glucosidic bond

9. **Bacterial cell wall contains which bond**
   (A) β 1-3 glycosidic bond
   (B) β 1-3 glucosidic bond
   (C) β 1-4 glucosidic bond
   (D) β 1-4 glycosodic bond

10. **Examples of Gram positive archaea containing pseudomurein are**
   (A) *Methanobacterium*
   (B) *Halobacterium*
   (C) *Thermobacterium*
   (D) *Halococcus*

11. Which Gram positive archaea do not contain pseudomurein

(A) *Halobacterium* and *Thermobacterium*

(B) *Methanosarcina* and *Halococcus*

(C) *Thermobacterium* and *Methanobacterium*

(D) *Thermobacterium* and *Halococcus*

12. Name Gram -ve archaea

(A) *Methanobacterium* and *Methanolobus*

(B) *Methanobacterium* and *Halobacterium*

(C) *Methanolobus* and *Halobacterium*

(D) *Methylococcus* and *Halobacterium*

13. Which Gram -ve archebacteria's cell wall contains glycoprotein

(A) *Methanococcus, Halobacterium* and *Desulfurococcus*

(B) *Methanomicrobium, Halobacterium* and *Desulfurococcus*

(C) *Methanolobus, Halobacterium* and *Sulfolobus*

(D) *Methanococcus, Halobacterium* and *Desulfurococcus*

14. Which G-ve archaeabacteria's cell wall contains protein?

(A) *Methanococcus, Methanomicrobium, Desulfurococcus*

(B) *Methanosarcinia, Methanobacterium, Desulfatomaculum*

(C) *Methanobacillus, Methanobacterium, Desulfatomaculum*

(D) *Methanococcus, Methanobacterium, Desulfatomaculum*

15. Most distinctive feature of archaea is

(A) Nature of their membrane lipid

(B) Nature of cell membrane

(C) Nature of their habitat

(D) Nature of their nucleus

16. Archaebacterial membrane lipids contains

(A) Fatty acids connected by ester links

(B) Branched chain hydrocarbons attached to glycerol by ether links

(C) Branched chain hydrocarbons attached to glycerol by ester links

(D) Fatty acids connected by ether links

17. Which type of linkage is found in archaeabacterial membrane lipids

(A) Ester linkage

(B) Glycosidic bond

(C) Peptide bond (D) Ether linkage

18. Archaeal genome is

(A) Single circular DNA

(B) Linear DNA

(C) Circular RNA (D) Linear RNA

19. Size of *Thermoplasma acidophilum* DNA is

(A) 2.5×10^9 daltons

(B) 0.8×10^9 daltons

(C) 1.8×10^9 daltons

(D) 2.5×10^8 daltons

20. Which archaeabacteria's genome is completely sequenced

(A) Methanobacterium jannaschii

(B) Methanospirillum jannaschii

(C) Methanobacillus jannaschii

(D) Methanococcus jannaschii

21. Splicing of RNA do not occur in

(A) Bacteria and Archaea

(B) Archaea and Fungi

(C) Bacteria and Yeast

(D) Bacteria and Fungi

22. **T C arm of archaeal t RNA contain**
(A) Uridine and 1-methyluridine
(B) Uridiline and 1-methyluridiline
(C) Pseudouridine and 1-methylpseudouridine
(D) Pseudouridiline and 1-methylpseudouridiline

23. **TψC arm of archaeal t RNA do not contain**
(A) Uracil (B) Adenine
(C) Guanine (D) Thymine

24. **Archaea are resistant to**
(A) Lysozyme
(B) β lactum antibiotics
(C) Both
(D) None

25. **Archaeal ribosomes are**
(A) 70 S (B) 80 S
(C) 40 S (D) 60 S

26. **Archaea contains which polymerase**
(A) RNA dependent DNA polymerase
(B) DNA dependent DNA polymerase
(C) DNA dependent RNA polymerase
(D) RNA dependent RNA polymerase

27. **Archaea do not contain enzyme**
(A) 6-phosphofructokinase and Pyruvate dehydrogenase complex
(B) Succinate dehydrogenase and 6-phosphofructokinase
(C) Succinate dehydrogenase and Pyruvate dehydrogenase complex
(D) Phosphatase and 6-phosphofructo-kinase

28. **Archaea are divided in to how many groups**
(A) Two (B) Three
(C) Four (D) Five

29. **Crenoarchaea are**
(A) Extreme thermophile
(B) Acidophilic
(C) Barophile
(D) Alkaliphilic

30. **Which archaea are strict anerobic and obtain energy by synthesizing methane**
(A) Methanotrophs
(B) Methanogens
(C) Methane oxidizers
(D) Methylotrophs

31. **Aerobic chemoheterotrophs that require at least 1.5 M NaCl for growth are known as**
(A) Moderate halophiles
(B) Halophiles
(C) Extreme halophile
(D) Hyperthermophile

32. **The microorganism (archaea) that can carry out photosynthesis using bacteriorhodopsin are**
(A) *Halobacterium salinarium*
(B) *Halomonas salinarium*
(C) *Halococcus salinarium*
(D) *Halospirillum salinarium*

33. **Which of the following is first amino acid in archaeal protein sysnthesis**
(A) Methionine
(B) Formylmethionin
(C) Proline
(D) Cystein

34. **Archaeal flagella are believed to be evolved from**
(A) Type II secretion system
(B) Type I secretion system
(C) Bacterial type IV pilli
(D) Bacterial type II secretion system

35. **Which of the following is true about archaebacteria**
(A) All halophiles
(B) All photosynthetic
(C) Old living entities
(D) All fossils

36. **Which of the following archebacteria is used for biogas production?**
(A) Methanogen
(B) Thermoacidophiles
(C) Halophiles
(D) Desulfurococcus

37. **Crenarchaeota comprise of how many classes**
(A) 2 (B) 3
(C) 1 (D) 7

38. **Pyrodictium belongs to**
(A) Crenarchaeota (B) Korarchaeota
(C) Nanoarchaeota (D) Euryarchaeota

39. **Euryarchaeota comprise of how many classes?**
(A) 4 (B) 5
(C) 3 (D) 7

40. **Methanogens belongs to**
(A) Crenarchaeota
(B) Korarchaeota
(C) Nanoarchaeota
(D) Euryarchaeota

41. **Largest group of archebacteria comprise of**
(A) Methanogens
(B) Hyperthermophiles
(C) Halophiles
(D) Thermophiles

42. **Cytoplasmic proteins and membrane lipids of extreme halophiles are**
(A) Acidic (B) Basic
(C) Neutral (D) Alkaline

43. ***Halobacterium salinarium* develops purple colour of membrane is due to**
(A) Bacteriochlorophyll
(B) Chlorophyll
(C) Bacteriorhodopsin
(D) Phycocyanin

44. **The archaea which contains cofactors F420 and methanopterin are**
(A) Methanogens, Sulphate reducing archaea and Archaeoglobus
(B) Methanotrophs and Sulphate reducing archaea
(C) Methanotrophs, Sulphate oxidizing archaea and Archaeoglobus
(D) Methanogens and Iron reducing archaea

| Answer Key |

Q. No.	Ans	Q. No.	Ans	Q. No.	Ans	Q. No.	Ans	Q. No.	Ans	Q. No.	Ans
1.	A	2.	A	3.	C	4.	B	5.	A	6.	D
7.	B	8.	A	9.	D	10.	A	11.	B	12.	C
13.	C	14.	A	15.	A	16.	B	17.	D	18.	A
19.	B	20.	D	21.	A	22.	C	23.	D	24.	C
25.	A	26.	C	27.	A	28.	A	29.	A	30.	B
31.	C	32.	A	33.	A	34.	C	35.	C	36.	A
37.	C	38.	A	39.	D	40.	D	41.	A	42.	A
43.	C	44.	A								

Chapter **5**

IMPORTANT GROUPS OF BACTERIA

1. **Which are the photosynthetic bacteria?**
   (A) Iron bacteria, Sulphur bacteria and Cyanobacteria
   (B) Purple bacteria, Green bacteria and Cyanobacteria
   (C) Cyanobacteria and Archebacteria
   (D) Iron bacteria and Green bacteria

2. **Which photosynthetic bacteria are able to carry out oxygenic photosynthesis**
   (A) Iron bacteria, Sulphur bacteria and Cyanobacteria
   (B) Cyanobacteria and Archebacteria
   (C) Iron bacteria and Green bacteria
   (D) Purple bacteria, Green bacteria and Cyanobacteria

3. **Which photosynthetic bacteria are able to carry out anoxygenic photosynthesis**
   (A) Cyanobacteria
   (B) Purple bacteria
   (C) Iron bacteria
   (D) Archebacteria

4. **The photosynthetic procedure in which bacteria use water as an electron donor and generate oxygen is known as**
   (A) Anoxygenic photosynthesis
   (B) Anerobic photosynthesis
   (C) Oxygenic photosynthesis
   (D) Oxidizing photosynthesis

5. **The photosynthetic bacteria which employ reduced molecules as their electron source for the generation of NADH and NADPH is known as**
   (A) Oxygenic photosynthesis
   (B) Anoxygenic photosynthesis
   (C) Aerobic photosynthesis
   (D) Reducing photosynthesis

6. **Purple and green photosynthetic bacteria contain which pigment?**
   (A) Chlorophyll
   (B) Bacteriochlorophyll
   (C) Bacteriorhodopsin
   (D) Phycocyanin

7. **Purple and green photosynthetic bacteria do not contain pigment**
   (A) Chlorophyll b
   (B) Bacteriochlorophyll
   (C) Phycocyanin
   (D) Chlorophyll a

8. **Cyanobacteria contain photosynthetic pigment**
   (A) Chlorophyll a
   (B) Chlorophyll b
   (C) Bacteriochlorophyll
   (D) Cyanophycin

9. **Chlorophyll absorbs which light ?**
   (A) Blue green (B) Red
   (C) Green (D) Blue and red

10. **Bacteriochlorophyll absorbs which light**
   (A) Visible light
   (B) Ultraviolet light
   (C) Infrared light
   (D) Invisible light

11. **Cyanobacteria contains which photosystem**
(A) Photosystem I and II
(B) Photosystem II and III
(C) Photosystem I and III
(D) Photosystem II and IV

12. **The accessory pigment present in cyanobacteria is**
(A) Phycocyanine
(B) Phycobilin
(C) Chlorophyll a
(D) Bacteriochlorophyll

13. **In cyanobacteria carbon dioxide (carbon) is assimilated by which cycle?**
(A) TCA cycle
(B) Oxidative phosphorylation
(C) Calvin cycle
(D) Banson cycle

14. **Blue green colour of cyanobacteria is due to presence of pigment**
(A) Phycocyanin
(B) Phycobilin
(C) Bacteriochlorophyll
(D) Chlorophyll a and b

15. **Red colour of cyanobacteria is due to presence of pigment**
(A) Bacteriochlorophyll
(B) Phycobilin
(C) Phycocyanin
(D) Phycoerythrene

16. **Specialized, dormant, thick-walled resting cells that are resistant to desiccation found in cyanobacteria are known as**
(A) Heterocyst (B) Akinites
(C) Afinites (D) Heterocells

17. **Nitrogen fixing cells in filamentous cyanobacteria is**
(A) Heterocyst (B) Akinites
(C) Photosome (D) Bacteroid

18. **The process in which light energy is trapped and converted to chemical energy is called**
(A) Respiration
(B) Catabolism
(C) Anabolism
(D) Photosynthesis

19. **Carotenoid is an**
(A) Essential pigment
(B) Accessory pigment
(C) Necessary pigment
(D) Potential pigment

20. **Green and purple bacteria uses which type of phosphorylation**
(A) Cyclic
(B) Acyclic
(C) Photogenic
(D) Aerobic

21. **Reaction center found in purple and green photosynthetic bacteria is**
(A) P420 (B) P580
(C) P870 (D) P470

22. **Metaquinone is present in which bacteria**
(A) Cyanobacteria, Purple and green sulfur bacteria
(B) Iron bacteria, Sulphur bacteria and Cyanobacteria
(C) Cyanobacteria and Archebacteria
(D) Iron bacteria and Green bacteria

23. **Endospores are commonly found in**
(A) G –ve bacetria
(B) G –ve archaea
(C) G + ve bacteria
(D) Gram variable bacteria

24. **Which genus contains spore forming bacteria**
(A) *Clostridium* and *Bacillus*
(B) *Azotobacter* and *Bacillus*
(C) *Chomatium* and *Azotobacter*
(D) *Cyanobacteria* and *Bacillus*

25. **Gram +ve anaerobic, spore forming bacteria that derives energy by sulphate reduction**
(A) Desulfobacter
(B) Desulfovibrio
(C) Desulfatomaculum
(D) Desulfolobus

26. **Cell wall less bacteria are known as**
(A) Rickettsia (B) Mycoplasma
(C) Phycoplasma (D) Archea

27. **Mycoplasma can not synthesize**
(A) Peptidoglycan (B) Lipid
(C) Protein (D) Nucleic acid

28. **Mycoplasma are**
(A) Gram positive
(B) Gram variable
(C) Gram negative
(D) Non reactive

29. **Mycoplasma belongs to which class**
(A) Mollicutes
(B) Firmicutes
(C) α-proteobacteria
(D) β-proteobacteria

30. **Smallest bacteria capable of self reproduction are**
(A) Spiroplasma (B) Nanobacteria
(C) Mycoplasma (D) Protoplast

31. **Mycoplasma exhibit which type of motility**
(A) Swimming (B) Running
(C) Zigzag (D) Gliding

32. **Mycoplasma are**
(A) Anaerobic
(B) Aerobic
(C) Microaerophilic
(D) Aerotolerant anaerobe

33. **Mycoplasma specifically requires which compound as nutrient**
(A) Sugar (B) Lipid
(C) Sterol (D) Vitamin

34. **Which biochemical pathway is not complete in mycoplasma?**
(A) Glycolysis
(B) ETC cycle
(C) HMP pathway
(D) TCA cycle

35. **Which of the following statements is true regarding the pathogenesis of Mycoplasma infections?**
(A) Enter the host cell by releasing exotoxin
(B) Adherence to the host cells mediated by the bacterial proteins
(C) Inhibited by beta-lactam antibiotics
(D) Enter the host cell throguh pili

36. **Which of the following statement do not hold true for Mycoplasma?**
(A) It can grow on cell-free media containing lipoprotein and sterol
(B) The size of the bacteria can be as small as 150 to 250 nm
(C) It can be identified using gram staining technique
(D) The bacteria show pleomorphic motility

37. **Which is the antibiotic producing actinomycetes ?**
(A) *Streptomyces* (B) *Nocardia*
(C) *Penicillium* (D) VAM

38. **Which one is the smallest autonomously replicating organism Virus**
(A) Viroid
(B) Mycoplasma
(C) Virus
(D) Actinomycetes

39. **Mycoplasma is**
(A) Unicellular eukaryote
(B) Multicellular eukaryote
(C) Unicellular prokaryote
(D) Multicellular prokaryote

40. The "witches broom" of legumes is caused by a

(A) Fungi (B) Bacteria
(C) Mycoplasma (D) Virus

41. Mycoplasmas are osmotically inactive and do not get affected by Vancomycin and Penicillin because they do not have

(A) Golgi body
(B) Mitochondria
(C) Nucleus
(D) Cell wall

42. Appearance of mycoplasma colonies is like

(A) Fungi (B) Fried egg
(C) Bread (D) Broom

43. The outermost layer in mycoplasma is

(A) Cell wall
(B) Cell membrane
(C) Sheath
(D) Capsule

44. The component which is responsible for the production of earthy smell by Streptomyces is

(A) Geojasmin
(B) Geosmin
(C) Streptomycin
(D) Erthocin

45. Agriculturally important actinomycetes is

(A) *Actinoallomurus*
(B) *Frankia*
(C) *Penicillium*
(D) *Allostreptomyces*

46. Nodule forming nitrogen fixing actiinomycetes found to colonize actinorhizal plants is

(A) *Frankia*
(B) *Streptomyces*
(C) *Actinoallomurus*
(D) *Penicillium*

47. Frankia follows which type of respiration

(A) Aerobic
(B) Anaerobic
(C) Microaerophilic
(D) Aerotolerant

48. *Streptococcus, Enterococcus, Lactococcus, Lactobacillus* and *Leuconostoc* are all members of which group of bacteria

(A) Acetic acid (B) Lactic acid
(C) Tartaric acid (D) Oxalic acid

49. Lactic acid bacteria obtain energy by

(A) Oxidative phosphorylation
(B) Photophosphorylation
(C) Substrate level phosphorylation
(D) Oxygenic phosphorylation

50. The organisms which can not synthesize vitamins, amino acids, purines, and pyrimidines and required them to be supplied in their medium are known as

(A) Fastidious organisms
(B) Autotrophs
(C) Auxotrophs
(D) Mixotrophs

51. Lactic acid bacteria are

(A) Autotroph (B) Auxotrophs
(C) Fastidious (D) Mixotrophs

52. Lactobacilli carry out homolactic fermentation by using which pathway

(A) HMP pathway
(B) TCA cycle
(C) ED pathway
(D) EMP pathway

53. Lactobacilli carry out heterolactic fermentation by using which pathway

(A) ED pathway
(B) Pentose phosphate pathway
(C) Glycolysis
(D) Glyoxalate pathway

54. ***Leuconostoc* carry out which type of fermentation?**
(A) Heterolactic (B) Homolactic
(C) None (D) Both

55. **Which of the following is halophilic bacteria**
(A) Halobacterium
(B) Dunaliella
(C) Ectothiorhodospira
(D) All

56. **Extreme halophile accumulate which component in their cells to become hypertonic**
(A) Potassium (B) Sodium
(C) Calcium (D) Chloride

57. **Halophilic bacterial cell wall and plasma membrane requires high concentration of which ion**
(A) Chloride (B) Calcium
(C) Sodium (D) Potassium

58. **Halophilic bacterial enzymes, ribosomes, and transport proteins requires presence of which ions for their optimum activity**
(A) Sodium (B) Chloride
(C) Potassium (D) Phosphate

59. **The extreme halophile which carry out photosynthesis is**
(A) *Halobacterium salinarium*
(B) Halobacterium salinitireducence
(C) Halobacillus salinarium
(D) Halobacillus salinitireducence

60. **Halophilic bacteria can survive in high salt concentration due to**
(A) Presence of branched chain lipids in cell membrane
(B) Presence of specialized resistant enzymes
(C) Presence of special lipids in cell membrane
(D) Presence of unbranched fatty acids in cell wall

61. **In photosynthetic halophile which pigment is found for trapping sunlight?**
(A) Bacteriochlorophyll
(B) Erythrosine
(C) Chlorophyll
(D) Bacteriorhodopsin

62. **The organisms which can grow at 55 °C are known as**
(A) Thermophiles
(B) Hypothermophiles
(C) Mesophiles
(D) Hyperthermophiles

63. **Examples of hyperthermophiles are**
(A) *Pyrococcus* and *Pyrodictium*
(B) *Sulfolobus*
(C) *Desulfurococcus*
(D) All

64. **In extreme thermophiles which is the reserve food material**
(A) Glucose
(B) Glycogen
(C) Galactose
(D) Sucrose

65. **Sulphur bacteria are**
(A) Neutrophilic
(B) Alkalophilic
(C) Acidophilic
(D) Alkali tolerant

66. **Thermophilic bacteria are**
(A) Thiobacillus
(B) Sulfolobus
(C) Thermoplasma
(D) All

67. **The organisms which are involved in formation of CH_4 are known as**
(A) Methanogens
(B) Methanotroph
(C) *Thiobacillus*
(D) *Thermococcus*

68. Methanogens are
(A) Obligate aerobes
(B) Facultative aerobe
(C) Obligate anaerobes
(D) Facultative anaerobe

69. Methanogens are classified as
(A) Archaea (B) Eubacteria
(C) Proteobacteria (D) Fungi

70. Which of the following component is found in methanogen cell wall
(A) Lipoprotein
(B) Glycoprotein
(C) Chaperone
(D) Osmoprotein

71. Which of the following is reserve food material in cyanobacteria
(A) Protein (B) Sugar
(C) Peptidoglycan (D) Glycogen

72. Pigment phycocyanin and phycoerythrin are found in
(A) Bacillariophyceae
(B) Archaebacteria
(C) Eubacteria
(D) Cynobacteria

73. Which pathway is not complete in methanogens
(A) PP pathway (B) ED pathway
(C) Glycolysis (D) TCA cycle

74. Methanogens can degrade
(A) Organic matter
(B) Inorganic matter
(C) Inorganic chemicals
(D) Phosphatic matter

| Answer Key |

Q. No.	Ans	Q. No.	Ans	Q. No.	Ans	Q. No.	Ans	Q. No.	Ans	Q. No.	Ans
1.	B	2.	D	3.	B	4.	C	5.	B	6.	B
7.	D	8.	A	9.	D	10.	C	11.	A	12.	B
13.	C	14.	A	15.	D	16.	B	17.	A	18.	D
19.	B	20.	A	21.	C	22.	A	23.	C	24.	A
25.	C	26.	B	27.	A	28.	C	29.	A	30.	C
31.	D	32.	A	33.	C	34.	D	35.	B	36.	C
37.	A	38.	B	39.	C	40.	C	41.	D	42.	B
43.	B	44.	B	45.	B	46.	A	47.	C	48.	B
49.	C	50.	A	51.	C	52.	D	53.	B	54.	A
55.	D	56.	A	57.	C	58.	C	59.	A	60.	C
61.	D	62.	A	63.	D	64.	B	65.	C	66.	D
67.	A	68.	C	69.	A	70.	B	71.	D	72.	D
73.	D	74.	A								

Chapter 6

VIRUSES

1. Virus particle consisting of DNA or RNA enclosed in protein coat are known as

(A) Virion (B) Viroid
(C) Prion (D) Mycoplasma

2. Which organisms are known as obligate intracellular parasites

(A) Fungi (B) Bacteria
(C) Archae (D) Virus

3. Protein coat of virus is known as

(A) Capsin (B) Capsid
(C) Capsomere (D) Cassin

4. Single stranded DNA viruses are

(A) Herpesviruses, T coliphages, Lambda phage
(B) Parvoviruses, ΦX174 phage, M13 phage
(C) Togaviruses, TMV, Most plant viruses
(D) Reoviruses, Wound-tumor virus of plants, Cytoplasmic

5. Which are double stranded DNA viruses

(A) Herpesviruses, T coliphages, Lambda phage
(B) Parvoviruses, ΦX174 phage, M13 phage
(C) Togaviruses, TMV, Most plant viruses
(D) Reoviruses, Wound-tumor virus of plants, Cytoplasmic

6. Which are single stranded RNA viruses?

(A) Herpesviruses, T coliphages, Lambda phage
(B) Parvoviruses, ΦX174 phage, M13 phage
(C) Togaviruses, TMV, Most plant viruses
(D) Reoviruses, Wound-tumor virus of plants, Cytoplasmic

7. Which are double stranded RNA viruses?

(A) Herpesviruses, T coliphages, Lambda phage
(B) Parvoviruses, ΦX174 phage, M13 phage
(C) Togaviruses, TMV, Most plant viruses
(D) Reoviruses, Wound-tumor virus of plants, Cytoplasmic

8. Envelop of enveloped viruses is made up of

(A) Nucleoproteins (B) Proteins
(C) Lipids (D) DNA

9. The virus or bacteriophage whose life cycle culminates (ends) with bursting of cell is known as

(A) Lysogenic cycle (B) Lytic cycle
(C) Cell cycle (D) Burst cycle

10. The bacterial viruses when gets integrated in to the bacterial genome and do not lyse the cells at the end of their life cycle is known as

(A) Lysogenic cycle
(B) Lytic cycle
(C) Cell cycle
(D) Burst cycle

11. The viruses which follow lytic cycle of growth are known as
(A) Temperate phage
(B) Burst phage
(C) Virulent phage
(D) Infectious phage

12. The viruses which follows lysogenic cycle are known as
(A) Virulent phage
(B) Temperate phage
(C) Infectious phage
(D) Lytic phage

13. The number of viruses produced per infected cell is known as
(A) Lytic size
(B) Lysogenic size
(C) Burst size
(D) Infection size

14. In bacteriophage, which structures helps attachment to host cell?
(A) Tail fibres
(B) Tail
(C) Capsid
(D) Capsomere

15. Bacteriophage capsid is made up of small protein subunits known as
(A) Capsomers
(B) Lipid droplets
(C) DNA fragments
(D) Polysachharide

16. In RNA virus, RNA replicates with the help of enzyme
(A) RNA dependent DNA poulemerase
(B) DNA dependent RNA polymerase
(C) DNA dependent DNA polymerase
(D) RNA dependent RNA polymerase

17. When the bacteriophage genome gets integrated in bacterial genome, the structure is known as
(A) Prophage (B) Epiphage
(C) Endophage (D) Exophage

18. The repressor protein responsible for conversion of lysogenic phage are
(A) Bro protein
(B) Cro proteins
(C) Kro protein
(D) Rec protein

19. Which of the life cycle of bacteriophage gets activated when cro protein level declines and γ repressor concentration increases
(A) Lytic cycle
(B) Lysogenic cycle
(C) Not dependent on concentration of cro protein
(D) cro protein concentration increases

20. Which life cycle occurs if cro protein level is high and γ repressor level is low
(A) Lytic cycle
(B) Lysogenic cycle
(C) Infective cycle
(D) Non-infective cycle

21. Repressor concentration is lowered down by activity of
(A) Rec C protein
(B) Rec B protein
(C) Rec A protein
(D) Rec D protein

22. Shape of γ repressor protein is
(A) Dumbel (B) Triangular
(C) Round (D) Rectangular

23. Active form of γ repressor is
(A) Monomeric (B) Dimeric
(C) Trimeric (D) Hexameric

24. dS DNA plant viruses
(A) Wound tumor, Rice dwarf virus
(B) Cauliflower mosaic virus and Dahlia mosaic virus
(C) Wound tumor virus and TMV
(D) Fiji disease virus and TMV

25. dS RNA plant viruses
(A) Wound tumor, Rice dwarf virus, Fiji disease virus
(B) Cauliflower mosaic virus, Dahlia mosaic virus, TMV
(C) Wound tumor, Dahlia mosaic virus, TMV
(D) Dahlia mosaic virus, Rice dwarf virus, Wound tumor

26. ss RNA plant viruses
(A) TMV
(B) Cauliflower mosaic virus
(C) Dahlia mosaic virus
(D) Wound tumor virus

27. Infection of plant with virus can be marked by formation of
(A) Yellowing of leaves
(B) Burning of plant
(C) Localized necrotic lesion
(D) Red patch on leaves

28. Within the plant, virus moves in which part of the plant
(A) Xylem (B) Periplasm
(C) Epiderm (D) Phloem

29. Which plant virus is transmitted by parasitic fungi?
(A) Tabacco necrosis virus
(B) Tobacco mosaic virus
(C) Cauliflower mosaic virus
(D) Wound tumor virus

30. Wound tumor virus can multiply in gut of which insect
(A) Grass hopper (B) Leaf hopper
(C) Earth warm (D) Ball warm

31. Which of the following virus infects alage
(A) *Coccolithovirus*
(B) *Phaeovirus*
(C) *Chlorovirus*
(D) All of above

32. The infectious particle containing only nucleic acid is known as
(A) Viroid (B) Virons
(C) Prions (D) Nuclear virus

33. The causative agent for potato spindle-tuber disease, exocortis disease of citrus trees, and chrysanthemum stunt disease is
(A) Prions (B) Virons
(C) Viroids (D) Plasmid

34. Protenaceous infectious particle are known as
(A) Capsid
(B) Nucleocapsid
(C) Prions
(D) Viroids

35. The first to isolate plant virus was
(A) W.M. Stanley
(B) E. C. Stackmann
(C) A. K. Smith
(D) Ivanowsky

36. Each capsomere of TMV contain amino acids whose number is
(A) 158 (B) 185
(C) 815 (D) 581

37. Which of the following is a DNA containing plant virus
(A) Tobacco mosaic virus
(B) Tomato mosaic virus
(C) Cauliflower mosaic virus
(D) Potato mosaic virus

38. Which of the following is called "filterable agent"
(A) Bacteria (B) Virus
(C) Fungi (D) Algae

39. Majority of plant viruses comprising of
(A) DNA
(B) RNA
(C) DNA and RNA both
(D) Lipids

40. **Viruses were called Contagium vivum fluidum by**
(A) de' Herelle
(B) Bawden and Pirie
(C) Twort
(D) Beijerinck

41. **Virusus usually infect the whole plant except the**
(A) Stem apex (B) Cortex
(C) Pith (D) Phloem

42. **Death of host cells due to viral infection is called**
(A) Stunting (B) Necrosis
(C) Rosetting (D) Wilting

43. **Shape of the tobacco mosaic virus is**
(A) Isometric (B) Rod
(C) Filamentous (D) Bacilliform

44. **Which of the virus has tadpole like shape**
(A) TMV
(B) DMV
(C) Human polio virus
(D) Bacteriophage

45. **The enzymes formed by bacteriophage which ruptures the bacterial wall is**
(A) Proenzymes
(B) Phage lysozymes
(C) Endolysins
(D) Phagoproteins

46. **The phage virus which has RNA as genetic material is**
(A) T2 phage (B) T4 phage
(C) T6 phage (D) MS2 phage

47. **Which of the following plant virus shows geminate shape?**
(A) Maize streak virus
(B) Tobacco mosaic virus
(C) Potato virus Y
(D) Cocoa swollen shoot virus

48. **In cyanophage, the genetic material is**
(A) RNA
(B) DNA
(C) RNA and DNA
(D) Protein

49. **When a virus enters a cell but does not replicate immediately, the situation is called**
(A) Lysogeny (B) Fermentation
(C) Symbiosis (D) Synergism

50. **Usually viruses are separated into several large groups based primarily on**
(A) Nature of the host
(B) Nucleic acid characteristics
(C) Capsid symmetry
(D) Diameter of the viroin or nucleocapsid

51. **The first step in infection of a host bacterial cells by a phage is**
(A) Adsorption (B) Absorption
(C) Penetration (D) Replication

52. **Viruses range in size from:**
(A) 1-100 nm (B) 25-300 nm
(C) 10-100 μm (D) 1-10 μm

53. **Bacteriophage are readily counted by the process of:**
(A) Immunoassays
(B) ELISA
(C) Plaque assays
(D) Electron Microscopy

54. **Which of the following is filamentous virus?**
(A) Maize streak virus
(B) Tobacco mosaic virus
(C) Potato virus Y
(D) Cocoa swollen shoot virus

55. **Prophage appears in**
(A) Lysogeny
(B) Lytic phase
(C) Spontaneous induction
(D) Spontaneous repression

56. A change from lysogeny to lytic is generally induced by

(A) Chemicals
(B) Irradiation
(C) Ultraviolet light
(D) Heat

57. The process in which ,the viral DNA is removed from the host's chromosomes and the lytic cycle occurs is known as

(A) Spontaneous induction
(B). Inductive infection
(C) Resultant induction
(D) Spontaneous infection

58. Which of the following is not present in Viroid?

(A) DNA (B) RNA
(C) Envelop (D) Cell membrane

59. The structure comprising of capsid protein and nucleic acid is known as

(A) Capsid (B) Capsomere
(C) Nucleocapsid (D) Nucleic acid

60. The virus containing host cell DNA instead of viral genome are known as

(A) Virions (B) Prions
(C) Bacteriophage (D) Pseudovirus

61. Which of the following plant viruses are transmitted through sap?

(A) Tobacco mosaic virus and Cucumber mosaic virus
(B) Tabacco necrosis virus and Wound tumor virus
(C) Tobacco mosaic virus and Tabacco necrosis virus
(D) Cucumber mosaic virus and Wound tumor virus

62. Which of the following plant viruses are transmitted by nematodes?

(A) Tabacco necrosis virus and Tobacco ringspot virus
(B) Tobacco mosaic virus and Cucmber mosaic virus
(C) Cucmber mosaic virus and Tobacco rattle virus
(D) Tobacco ringspot virus and Tobacco rattle virus

63. Shape of the Tobacco necrosis virus is

(A) Isometric (B) Rod shaped
(C) Filamentous (D) Geminate

64. The name corona virus was due to

(A) Crown-like projections
(B) Leaf-like projections
(C) Surface structure of bricks
(D) Round shape

65. The first case of novel coronavirus was found in

(A) Beijing (B) Shanghai
(C) Wuhan (D) Tianjin

66. Corona virus posses which type of genome?

(A) SS RNA (B) DS RNA
(C) SS DNA (D) DS DNA

67. The virus which cause Covid-19 is known as

(A) MERS-CoV-2
(B) SARS-CoV-2
(C) MERS-CoV-1
(D) SARS-CoV-1

68. Kuru disease is caused by

(A) Prion (B) Viroids
(C) Virons (D) Bacteriophage

| Answer Key |

Q. No.	Ans	Q. No.	Ans	Q. No.	Ans	Q. No.	Ans	Q. No.	Ans	Q. No.	Ans
1.	A	2.	D	3.	B	4.	B	5.	A	6.	C
7.	D	8.	B	9.	B	10.	A	11.	C	12.	B
13.	C	14.	A	15.	A	16.	D	17.	A	18.	B
19.	B	20.	A	21.	C	22.	A	23.	B	24.	B
25.	A	26.	A	27.	C	28.	D	29.	A	30.	B
31.	D	32.	A	33.	C	34.	C	35.	A	36.	A
37.	C	38.	B	39.	B	40.	D	41.	A	42.	B
43.	B	44.	D	45.	C	46.	D	47.	A	48.	B
49.	A	50.	A	51.	A	52.	B	53.	C	54.	C
55.	A	56.	C	57.	A	58.	C	59.	C	60.	D
61.	A	62.	D	63.	A	64.	A	65.	C	66.	A
67.	B	68.	A								

Chapter 7

SOIL MICROBIAL ECOSYSTEM-RHIZOSPHERE AND PHYLLOSPHERE

1. **The natural place where microbes or their community resides is known as**
   (A) Habitat (B) Biosphere
   (C) Niche (D) Biome

2. **Which one of the following is characteristic of ideal ecosystem?**
   (A) Primary consumers are less dependent on producers
   (B) Primary consumers are more than producers
   (C) Producers are more than primary consumers
   (D) Secondary consumers are more and powerful

3. **In an ecosystem which of the following shows unidirectional flow**
   (A) Carbon (B) Nitrogen
   (C) Potassium (D) Free energy

4. **Which of the following term define soil conservation?**
   (A) Aeration of soil
   (B) Conversion of sterile soil into fertile one
   (C) Soil erosion
   (D) Protection against loss

5. **As per Shelford's low of tolerance, microbes with wide tolerance for various environmental factors have**
   (A) Wide distribution with low population size
   (B) Wide distribution with high population size
   (C) Narrow distribution with low population size
   (D) Narrow distribution with high population size

6. **Two or more populations of the organism living in an environment is known as**
   (A) Niche (B) Community
   (C) Biome (D) Habitat

7. **Wide variety of organisms are known as**
   (A) Biodiversity (B) Habitat
   (C) Population (D) Niche

8. **The volume of soil that surrounds a germinating seed is known as**
   (A) Rhizosphere (B) Rhizoplane
   (C) Spermosphere (D) Phyllosphere

9. **The region where soil and roots of the plants make contact or the thin layer of soil adhering to a root system after shaking and removing the loose soil is known as**

   Or

   The soil closely adhered to plant roots is known as

   Or

   The volume of soil around the root influenced by the materials released from the (root exhudates) is known as
   (A) Rhizosphere
   (B) Rhizoplane
   (C) Bulk soil
   (D) Phyllosphere

10. The direct influence of plant roots on microbes and microbes on plant roots within the Rhizosphere is known as

(A) Root exudate effect
(B) Rhizosphere effect
(C) Microbial efficiency
(D) Rhizoplane effect

11. The sum total of the populations of the same kind of organisms constitute

(A) Genus (B) Community
(C) Species (D) Colony

12. The abundance of a species population, within its habitat, is called

(A) Relative density
(B) Habitat density
(C) Niche density
(D) Absolute density

13. Initially during early stage of plant life which type of compounds are produced by root exhudate

(A) Lipids
(B) Proteins
(C) Vitamins
(D) Carbohydrates

14. At maturation of roots which type of compound are released in soil

(A) Simple sugars
(B) Complex proteins
(C) Fatty acids
(D) Nucleic acids

15. In rhizosphere which type of bacteria found in high proportion

(A) Gram +ve cocci
(B) Gram –ve rods
(C) Gram +ve rods
(D) Gram variable

16. Which are the toxic compounds found in root exudates

(A) Glycosides (B) HCl
(C) Cyanide (D) H_2O_2

17. The growth hormones secreted by microorganisms which stimulates root hair development in plants are

(A) Auxins
(B) Jasmonic acid
(C) Cytokinin
(D) Salicylic acid

18. The compound produced in rice field which is toxic to the plant roots is

(A) Hydrogen sulphide
(B) Hydrogen cyanide
(C) Hydrogen chloride
(D) Hydrogen phosphate

19. Hydrogen sulphide in rice field is neutralized by which microorganism

(A) *Beggiatoa*-sulphur reducing bacteria
(B) *Azotobacter*-nitrogen fixing bacteria
(C) *Beggiatoa*- sulphur oxidizing bacteria
(D) *Azospirillum*-nitrogen fixing bacteria

20. When the substances produced by microorganisms or roots can positively or negatively affect plant and microbial growth than this effect is known as

(A) Allelopathic effect
(B) Allelotrophic effect
(C) Exotrophic effect
(D) Endotrophic effect

21. Which of the following term defines closely related, morphologically similar, sympatric populations which are reproductively isolated?

(A) Clines (B) Sibling species
(C) Denies (D) Clones

22. In an arable soil bacterial content in soil is

(A) 10-1000 kg/ha (B) 30-3000 kg/ha
(C) 10-1000 g/ha (D) 30-3000 g/ha

23. In an arable soil actinomycetes content in soil is
(A) 40-400 kg/ha
(B) 400-4000 g/ha
(C) 400-4000 kg/ha
(D) 400-4000 t/ha

24. In an arable soil fungal content in soil is
(A) 40-5000 kg/ha
(B) 500-5000 g/ha
(C) 500-5000 kg/ha
(D) 500-5000 t/ha

25. In soil which microorganisms are found in higher content
(A) Fungi
(B) Virus
(C) Actinomycetes
(D) Bacteria

26. Zone of phyllosphere surrounding fruits is known as
(A) Anthosphere (B) Carposphere
(C) Phylloplane (D) Caulosphere

27. C:N ratio of microorganisms
(A) 4-9:1 (B) 5-8:2
(C) 4-9:2 (D) 5-8:1

28. Native soil bacteria are also known as
(A) Autocthonus species
(B) Allochthonous species
(C) Alien species
(D) Autogrowing species

29. The bacteria which alien or artificially invading the soil are known as
(A) Allocthonus species
(B) Native species
(C) Autochthonous
(D) Autogrowing species

30. The physical structure of soil is improved mainly by the accumulation of
(A) Fungus (B) Bacteria
(C) Viruses (D) Protozoa

31. Ratio of rhizospheric soil microorganisms : bulk soil microorganisms is known as
(A) R:B ratio (B) R:M ratio
(C) R:R ratio (D) R:S ratio

32. Methylotroph utilize which component of plant exudates?
(A) Methanol
(B) Methane
(C) Methyl ammonium sulphate
(D) Methyl bromide

33. Methanol on plant leaves is produced as bi-product of
(A) Cellulose degradation
(B) Hemicellulose degradation
(C) Chitin degradation
(D) Lignin degradation

34. Who first time coined the term rhizosphere
(A) Windogradsky
(B) Lorenz Hiltner
(C) Beijerinck
(D) D' Bary

35. The phyllospheric habitat surrounding stem is known as
(A) Anthosphere (B) Carposphere
(C) Phylloplane (D) Caulosphere

36. The surface of plant roots is known as
(A) Rhizoplane
(B) Rhizosphere
(C) Bulk soil
(D) Spermatosphere

37. Bacterial population in surface soil is
(A) 10^7-10^8 cells / g dry weight of soil
(B) 10^7-10^8 cells / kg dry weight of soil
(C) 10^8-10^9 cells / g dry weight of soil
(D) 10^8-10^9 cells / kg dry weight of soil

38. How much percentage of soil microorganisms are culturable
(A) 90 % (B) 70%
(C) 50 % (D) 10 %

39. The phenomenon of ice nucleation is seen in

(A) Bacteria (B) Viruse
(C) Fungi (D) Actinomycetes

40. Name of ice nucleating bacteria is

(A) *Bacillus syringae*
(B) *Pseudomonas putida*
(C) *Bacillus coagulans*
(D) *Pseudomonas syringae*

41. The phenomenon of bacteria where in it synthesize special proteins which act as nucleation centre for snow formation in cloud chamber is known as

(A) Water nucleation
(B) Nucleation
(C) Ice nucleation
(D) Frost nucleation

42. Zone of phyllosphere surrounding leaves is known as

(A) Anthosphere (B) Carposphere
(C) Phylloplane (D) Caulosphere

43. Which of the following is indicator of rhizosphere inhibition?

(A) R:S ratio ≥ 1
(B) R:S ratio < 1
(C) R:S ratio = 1
(D) R: N ratio =1

44. The bacteria in whose presence the water can not be freezed readily on plant leaf surfaces are known as

(A) Ice nucleation bacteria
(B) Ice resistance bacteria
(C) Ice minus bacteria
(D) Ice tolerant bacteria

45. In soil methane gas is produced by activity of which bacteria

(A) Methanotrophs
(B) Methanogens
(C) Methane utilizers
(D) Methane oxidizers

46. Methanogens are

(A) Microaerophillic bacteria
(B) Aerobic bacteria
(C) Anaerobic bacteria
(D) Obligate aerobic bacteria

47. Degradation of organic matter by methanogens in absence of oxygen results in formation of

(A) Methane gas
(B) Carbon dioxide
(C) Hydrogen
(D) Sulphur dioxide

48. Which organisms can utilize Methane gas

(A) Methanotrophs
(B) Methanogen
(C) Acetogens
(D) Acylotrophs

49. The phyllospheric zone surrounding flowers is known as

(A) Anthosphere (B) Carposphere
(C) Phylloplane (D) Caulosphere

50. Methanotrophic bacteria are

(A) Anaerobic
(B) Aerobic
(C) Microaerophilic
(D) Aerotolerant anaerobe

51. Detritivores are best defined as the organisms that ....

(A) Can live without oxygen
(B) Do not need to consume any organic carbon
(C) Require oxygen to live
(D) Consume dead remains of other organisms

52. Rhizosphere effect is almost negligible on which of the following organisms?

(A) Bacteria (B) Fungi
(C) Actinomycetes
(D) Algae

53. Which of the following type of microorganisms are having maximum rhizospheric effect

(A) Bacteria (B) Fungi
(C) Algae (D) Protozoa

54. Rhizospheres affect microbial population and nutrient availability by

(A) Exerting physical pressure
(B) Secreting nutritive metabolites and organic solvents
(C) Increasing pore space
(D) Changing soil structure

55. As the temperature and light intensity increase the number of rhizospheric microorganisms tend to

(A) Increase (B) Decrease
(C) Not affected (D) First decrease and than increase

56. Which of the following is indicator of good rhizosphere stimulation?

(A) R:S ratio > 1 (B) R:S ratio < 1
(C) R:S ratio = 1 (D) R:S ratio ≤ 1

57. Acidic soil serves as favourable habitat for

(A) Bacteria (B) Fungi
(C) Actinomycetes (D) Protozoa

58. Phyllospheric microorganisms are generally termed as

(A) Endophyte (B) Exophyte
(C) Epiphyte (D) Phyllophyte

59. As the distance from the roots increases population of bacteria tend to

(A) Increase
(B) Decrease
(C) Not affected
(D) Remains constant

60. Soil microbial biomass can be measured by

(A) GC-MS method
(B) Chloroform method
(C) ICP-MS method
(D) DTPA extraction method

| Answer Key |

Q. No.	Ans	Q. No.	Ans	Q. No.	Ans	Q. No.	Ans	Q. No.	Ans	Q. No.	Ans
1.	A	2.	C	3.	D	4.	D	5.	B	6.	B
7.	A	8.	C	9.	A	10.	B	11.	C	12.	C
13.	D	14.	A	15.	B	16.	A	17.	A	18.	A
19.	C	20.	A	21.	D	22.	B	23.	C	24.	C
25.	D	26.	B	27.	A	28.	B	29.	C	30.	A
31.	D	32.	A	33.	D	34.	B	35.	D	36.	B
37.	C	38.	D	39.	A	40.	D	41.	C	42.	C
43.	C	44.	C	45.	B	46.	C	47.	A	48.	A
49.	A	50.	B	51.	D	52.	D	53.	A	54.	B
55.	A	56.	A	57.	B	58.	C	59.	B	60.	B

Chapter **8**

MICROBIOLOGY OF WATER: FRESH, MARINE AND WASTE

1. The microorganism isolated from thermophilic habitat of yellow stone national park which has found tremendous application in molecular biology

(A) *Thermus aquaticus*
(B) *Thermospora aquaticus*
(C) *Thermococcus aquaticus*
(D) *Bacillus megaterium*

2. Most of the available antibiotics are produced by

(A) Protozoa (B) Yeast
(C) Actinomycetes (D) Virus

3. Nutrient poor lake of fresh water is known as

(A) Eutrophic lake
(B) Oligotrophic lake
(C) Polytrophic lake
(D) Mal nutrition lake

4. Nutrient rich lake of fresh water is known as

(A) Nutrient rich lake
(B) Oligotrophic lake
(C) Polytrophic lake
(D) Eutrophic lake

5. Which type of lake is oxygen saturated and poor in microbial population

(A) Eutrophic lake
(B) Oligotrophic lake
(C) Polytrophic lake
(D) Photic lake

6. Which type of lake contains bottom sediments of organic matter

(A) Oligotrophic lake
(B) Polytrophic lake
(C) Eutrophic lake
(D) Photic lake

7. The upper warm and aerobic layer in eutrophic lake is known as

(A) Epilimnion
(B) Endolimnion
(C) Hypolimnion
(D) Hyperlimnion

8. The dipper colder bottom anaerobic layer in eutrophic lake is known as

(A) Hyperlimnion
(B) Epilimnion
(C) Endolimnion
(D) Hypolimnion

9. When the large amount of nutrients are added in to the water the process is known as

(A) Eutrophication
(B) Hypertrophication
(C) Hyperaccumulation
(D) Hypotrophication

10. The massive growth of algae and cyanobacteria in surface water is known as

(A) Gloom
(B) Mat
(C) Bloom
(D) Floccus

11. **When the phosphorous content of water is high than which process occurs**
(A) Massive algal growth
(B) Massive bacterial growth
(C) Massive fungal growth
(D) Massive virus growth

12. **Which type of microorganisms are found in anaerobic layer (hypolimnion) of eutrophic lake**
(A) Photosynthetic iron microorganisms
(B) Chemotrophic sulphur microorganisms
(C) Photosynthetic sulphur microorganisms
(D) Chemotrophic iron microorganisms

13. **The algae which is freely floating in water are known as**

Or

A community of floating photosynthetic organisms, largely composed of algae and cyanobacteria
(A) Photoplanktone
(B) Zooplanktone
(C) Phytoplanktone
(D) Zygoplanktone

14. **A community of floating, aquatic, minute animals and non-photosynthetic protists**
(A) Phytoplankton
(B) Zooplankton
(C) Photoplanktones
(D) All of above

15. **The carbon-nitrogen-phosphorus (C:N:P) ratio of aquatic microorganisms (phytoplankton)**
(A) Redfield ratio
(B) Brownfield ratio
(C) Greenfield ratio
(D) Bluefield ration

16. **The term which predict limiting factor for microbial growth in aquatic habitat is**
(A) Brownfield ratio
(B) Redfield ratio
(C) Greenfield ratio
(D) Bluefield ration

17. **Commonly used value of redfield ratio is**
(A) 16C:16N:1P (B) 16C:106N:1P
(C) 16C:16N:16P (D) 106C:16N:1P

18. **Cyanobacteria works best at which pH**
(A) Acidic (4.5-6.5)
(B) Neutral (7)
(C) Acidic (1-4)
(D) Alkaline (8.5-9.5)

19. **Algae works best at which pH**
(A) Acidic (B) Alkaline
(C) Neutral (D) Extreme acidic

20. **Top producers in aquatic habitat are**
(A) Phytoplanktons
(B) Aquatic weeds
(C) Aquatic ferns
(D) Aquatic plants

21. **Phytoplanktons that are smaller than 2 µm are known as**
(A) Microplanktons
(B) Nanoplanktons
(C) Miliplanktons
(D) Picoplanktons

22. **1/3 of the oceanic picoplankton (cells smaller than 2 µm) are**
(A) Bacteria (B) Actinobacteria
(C) Algae (D) Archaea

23. **The fungi which is involved in recycling of leaves (organic matter) in aquatic habitat and are**
(A) Ingoldian
(B) Engloindian
(C) Egoldian
(D) Englodians

24. The bacteria which can grow between 0 and 400 atm, but best at normal atmospheric pressure are known as
(A) Barophilic
(B) Barotrophic
(C) Barotolerant
(D) Barogenic

25. The bacteria which prefer higher pressure are known as
(A) Barophilic
(B) Barotolernt
(C) Barotrophic
(D) Barogenic

26. Most of the nutrient cycling and photosynthesis occurs in upper 300 m zone of water known as
(A) Benthic zone (B) Limnetic zone
(C) Littoral zone (D) Photic zone

27. Which type of microorganisms utilize methane present in methane hydrates
(A) Archaea
(B) Psychrophilic bacteria
(C) Psychrotrophic bacteria
(D) Eubacteria

28. When microorganisms metabolize methane deposits at low hydrogen levels under sulfate reducing conditions in ocean the process is known as
(A) Reverse sulfurylation
(B) Reverse methane oxidation
(C) Reverse methanotrophism
(D) Reverse methanogenesis

29. The problem which is occurring due to pollution of water are
(A) Algal blooms and Red tides
(B) Cyanobacterial blooms and Blue tides
(C) Phtobacterial blooms and green tides
(D) Phtobacterial blooms and Blue tides

30. Algal blooms are produced by which type of organisms
(A) Anabaena, Nostoc, Cylindrospermum
(B) Gymnodinium, Gonyaulax, Oscillatoria
(C) Gymnodinium, Gonyaulax and Cylindrospermum
(D) Anabaena, Gymnodinium, Gonyaulax

31. Red tides are producd by
(A) Anabaena, Nostoc, Cylindrospermum
(B) Gymnodinium, Gonyaulax, Oscillatoria
(C) Gymnodinium, Gonyaulax and Cylindrospermum
(D) Anabaena, Gymnodinium, Gonyaulax

32. Red tide microorganism produce which neurotoxin
(A) Saxitoxin (B) Maxitoxin
(C) Loxitoxin (D) Brevitoxin

33. The zone of water body near shore is known as
(A) Limnetic zone (B) Benthic zone
(C) Littoral zone (D) Profundal zone

34. The zone of open water away from shore is known as
(A) Profundal zone (B) Limnetic zone
(C) Benthic zone (D) Littoral zone

35. The zone of deeper water under the limnetic zone is known as
(A) Benthic zone (B) Littoral zone
(C) Limnetic zone (D) Profundal zone

36. The sediment at the bottom of water body is known as
(A) Profundal zone
(B) Limnetic zone
(C) Littoral zone
(D) Benthic zone

37. In profundal zone which type of microorganisms are seen

(A) Anaerobic bacteria

(B) Green photosynthetic bacteria

(C) Purple photosynthetic bacteria

(D) All

38. The microorganisms from lakes and rivers can grow at a salt concentration of

(A) 2.5 to 4 % (B) 5 percent

(C) < 1 percent (D) > 1 percent

39. The coliform group of bacteria includes all the ________________ bacilli.

(A) Aerobic

(B) Anaerobic

(C) Aerobic, non-sporulating

(D) Aerobic, gram-negative, non-sporulating

40. Which of the following organism belong to coliform group of bacteria?

(A) *Salmonella* (B) *Serratia*

(C) *Enterobacter* (D) *Proteus*

41. Among the following which microorganism acts as a representative of fecal streptococci group?

(A) *Streptococcus faecalis*

(B) *Streptococcus bovis*

(C) *Streptococcus equinus*

(D) *Streptococcus faecium*

42. Drifting or floating organisms found in the water body are known as

(A) Limnos (B) Benthos

(C) Weedos (D) Planktons

43. Rooted vegetation is found in which zone of water body?

(A) Littoral zone (B) Limnetic zone

(C) Profundal zone (D) Benthic zone

44. In regions of the estuary which of the following microorganisms found abundantly?

(A) Coliforms

(B) Appendaged bacteria

(C) Viruses

(D) Fecal streptococci

45. Red sea is having characteristic colour due to presence of

(A) Diatoms

(B) Cyanobacteria

(C) Dinoflagellates

(D) Chlamydomonads

46. Which of the following genus of bacteria is not found in fresh waters?

(A) *Pseudomonas*

(B) *Flavobacterium*

(C) *Aeromonas*

(D) *Vibrio*

47. Microbes present at the bottom of the ocean are known as

(A) Benthos

(B) Limnos

(C) Planktons

(D) Phytoplanktons

48. In benthic zone which type of microorganism found

(A) *Desulfovibrio*

(B) Methane bacteria

(C) *Clostridium*

(D) All

49. Which of the following treatment involve oxidation of organic constituents of the wastewater?

(A) Primary treatment

(B) Secondary treatment

(C) Advanced treatment

(D) Tertiary treatment

50. In activated sludge aeration period is about

(A) 1 hour

(B) 24 hours

(C) 10-15 hours

(D) 4-8 hours

51. Trickling filter is used in which of the following wastewater treatment processes?

(A) Primary treatment
(B) Secondary treatment
(C) Tertiary treatment
(D) Final treatment

52. The upper region of the trickling filter favours growth of

(A) Fungi (B) Protozoa
(C) Algae (D) Bacteria

53. Advanced treatment processes generally used to treat waste water to

(A) Remove coarse solids
(B) Remove settleable solids
(C) Reduce BOD
(D) Remove additional objectionable substances

54. Which of the following is correct sequence for treatment of municipal water supplies

(A) Coagulation, filtration, chlorination
(B) Chlorination, filtration, coagulation
(C) Filtration, coagulation, chlorination
(D) Coagulation, chlorination, filtration

55. Aggregation of microbial masses in activated sludge is known as

(A) Activated sludge
(B) Masses
(C) Colloids (D) Floccules

56. During secondary waste water treatment, oxidation ponds promotes the growth of

(A) Fungi (B) Bacteria
(C) Plant (D) Algae

57. Sedimentation of large particulate matter is enhanced by

(A) Aluminium sulphate
(B) Potassium nitrate
(C) Potassium permanganate
(D) Chlorination

58. In sewage water, BOD (biological oxygen demand) is primarily used for Estimating the

(A) Types of microbes
(B) Level of dissolved oxygen
(C) Quantity of organic matter
(D) Types of chemicals

59. The biggest advantage of a pure oxygen activated sludge process is:

(A) Requiring less tank volume
(B) Improving the nitrification process
(C) Reducing chlorine demand
(D) Requiring less mechanical equipment

60. The main purpose of covering trickling filters is to prevent :

(A) Algal growth on the media
(B) Odours
(C) Growth of flies and snails
(D) Heat loss and increase efficiency

61. A low respiration rate value (less than 4 mg O_2/hr/gm MLSS) indicates:

(A) Rapidly biodegradable organic matter
(B) Over-stabilized organic matter
(C) Normal rate for conventional activated sludge process
(D) Typical rate for high rate processes

62. The Redfield ratio is an index of concentration of

(A) Carbon, hydrogen and oxygen
(B) Nitrogen, potassium and iron
(C) Carbon, nitrogen and phosphorous
(D) Carbon, phosphorous and sulfur

63. Chemical precipitation of phosphorus is

(A) Primary treatment
(B) Secondary treatment
(C) Tertiary treatment
(D) Advanced treatment

64. Which of the following bacterial genera are not found in a maturing Winogradsky column except

(A) *Clostridium*
(B) *Rhodospirillum*
(C) *Chlorobium*
(D) *Escherichia*

65. Trickling filter and activated sludge are involved in

(A) Primary treatment
(B) Secondary treatment
(C) Tertiary treatment
(D) Advanced treatment

66. The fresh water ecosystem becomes eutrophic when

(A) Rate of respiration exceed that of photosynthesis
(B) Rate of respiration do not exceed that of photosynthesis
(C) Rate of photosynthesis exceed that of respiration
(D) Rate of photosynthesis and respiration are equal

67. Which of the following cyanobacterial genera carry out nitrogen fixation under anoxic condition using hydrogen sulphide as electron donor for photosynthesis?

(A) *Anabaena*
(B) *Nostoc*
(C) *Cylindrospermum*
(D) *Oscillatoria*

68. The pH and temperature more suitable for activity of cyanobacteria in water is

(A) pH-6.5-7.5; Temperature 25-28°C
(B) pH-6.5-7.5; Temperature 38-40°C
(C) pH-8.5-9.5; Temperature 38-45°C
(D) pH-8.5-9.5; Temperature 30-35°C

69. Which of the following system keeps the oceans buffered at pH 7.6 to 8.3

(A) Carbonate equilibrium system
(B) Hydroxide equilibrium system
(C) Hydrogen equilibrium system
(D) Nitrate equilibrium system

70. Which of the following marine bacteria are found in the coastal areas where nutrient mixing occurs

(A) *Thioploca* and *Thiomargarita*
(B) *Thiothrix and Thiobacillus*
(C) *Thiothrix and Thioploca*
(D) *Thiobacillus and Thiomargarita*

71. Which of the following is the world's largest known bacterium found in aquatic environment

(A) *Thiomargarita* (B) *Thioploca*
(C) *Sphaerotilus* (D) *Leucothrix*

72. Which of the following are aerobic gliding bacteria?

(A) *Sphaerotilus* and *Leucothrix*
*(B) Caulobacte*r and *Hyphomicrobium*
(C) *Flexithrix* and *Flexibacter*
(D) *Sphaerotilus* and *Hyphomicrobium*

73. Which of the following is filamentous nitrogen fixing cyanobacterium found in the ocean

(A) Trichodesmium
(B) Chlorobium
(C) Chgromatium
(D) Thiomargarita

74. Which of the following is most common and preferred method for microbiological analysis of water quality?

(A) Most probable number
(B) Presence-absence (P-A) test
(C) Defined substrate test
(D) Membrane filtration technique

75. BOD removal capacity of trickling filter is around

(A) 50-60 % (B) 30-40%
(C) 80-85 % (D) 60-70 %

| Answer Key |

Q. No.	Ans	Q. No.	Ans	Q. No.	Ans	Q. No.	Ans	Q. No.	Ans	Q. No.	Ans
1.	A	2.	C	3.	B	4.	D	5.	B	6.	C
7.	A	8.	D	9.	A	10.	C	11.	A	12.	C
13.	C	14.	B	15.	A	16.	B	17.	D	18.	D
19.	C	20.	A	21.	D	22.	D	23.	A	24.	C
25.	A	26.	D	27.	A	28.	D	29.	A	30.	A
31.	B	32.	A	33.	C	34.	B	35.	D	36.	D
37.	D	38.	C	39.	D	40.	C	41.	A	42.	D
43.	A	44.	B	45.	B	46.	D	47.	A	48.	D
49.	B	50.	D	51.	B	52.	C	53.	D	54.	A
55.	D	56.	D	57.	A	58.	C	59.	A	60.	D
61.	D	62.	C	63.	C	64.	D	65.	B	66.	A
67.	D	68.	D	69.	A	70.	A	71.	A	72.	C
73.	A	74.	D	75.	C						

◈

Chapter 9

MICROBIAL GROWTH CURVE AND MATHEMATICAL EXPRESSION OF GROWTH

1. **During culturing bacteria when the nutrients are provided once at the beginning of fermentation cycle, the type of culture is known as**
   (A) Batch culture
   (B) Fed batch culture
   (C) Continuous culture
   (D) Synchronous culture

2. **Batch culture is known as**
   (A) Open system
   (B) Diauxic system
   (C) Closed system
   (D) Circulated system

3. **Microbial growth curve contains how many phases**
   (A) One (B) Two
   (C) Three (D) Four

4. **The first phase of growth curve is_**
   (A) Lag phase
   (B) Exponential phase
   (C) Stationary phase
   (D) Decline phase

5. **In which phase of growth curve the cell is synthesizing new cell components**
   (A) Log phase
   (B) Stationary phase
   (C) Lag phase
   (D) Exponential phase

6. **In which phase of growth curve the microorganism get acclimatized or adapted to culture conditions**
   (A) Log phase
   (B) Lag phase
   (C) Stationary phase
   (D) Death phase

7. **When the inoculum is from old culture or refrigerated culture than length of lag phase is**
   (A) Short (B) Medium
   (C) Long (D) Absent

8. **When the young exponential growth phase culture is used as inoculum than the lag phase is**
   (A) Short (B) Absent
   (C) Long (D) Longest

9. **The organisms are growing and dividing at the maximum possible rate that phase of growth curve is known as**
   (A) Lag phase
   (B) Exponential phase
   (C) Stationary phase
   (D) Decline phase

10. **The rate of growth of microorganism at exponential phase is dependent on**
   (A) Genetic potential of culture
   (B) Nature of medium
   (C) Cultural conditions
   (D) All

11. **The rate of growth of microorganism in exponential phase is**
   (A) Increasing (B) Constant
   (C) Decreasing (D) Fluctuating

12. **In which phase of growth curve, the microorganism population is more uniform in terms of chemistry and physiology**
(A) Lag phase (B) Log phase
(C) Stationary phase
(D) Decline phase

13. **The exponential phase cultures are generally used for which studies**
(A) Nutrient use efficiency
(B) Waste accumulation efficiency
(C) Secondary metabolite production efficiency
(D) Biochemical and physiological characterization

14. **When all cellular constituents are manufactured at constant rates relative to each other the type of growth is known as**
(A) Synchronous growth
(B) Diauxic growth
(C) Balanced growth
(D) Unbalanced growth

15. **Microbial growth in exponential or log phase is**
(A) Synchronous growth
(B) Balanced growth
(C) Diauxic growth
(D) Unbalanced growth

16. **The increase in bacterial growth during exponential phase is in**
(A) Geometric progression
(B) Cyclic progression
(C) Regressive progression
(D) Binary progression

17. **Growth during which the rates of synthesis of cell components vary relative to one another is known as**
(A) Synchronous growth
(B) Balanced growth
(C) Diauxic growth
(D) Unbalanced growth

18. **The phase of growth in which population growth ceases and the growth curve becomes horizontal is known as**
(A) Log phase
(B) Stationary phase
(C) Stagnant phase
(D) Lag phase

19. **Stationary phase for bacteria is generally reached at population density of**
(A) 10^7 cells/ml
(B) 10^8 cells/ml
(C) 10^9 cells/ml
(D) 10^{10} cells/ml

20. **Stationary phase for algae and protozoa is reached at population density of**
(A) 10^8 cells/ml (B) 10^7 cells/ml
(C) 10^6 cells/ml (D) 10^5 cells/ml

21. **In which phase of growth curve total number of viable microorganisms remains constant**
(A) Stationary phase
(B) Lag phase
(C) Exponential phase
(D) Decline phase

22. **When the nutrients get limited and waste material is accumulated than the microbial population enters in to**
(A) Log phase
(B) Exponential phase
(C) Stationary phase
(D) Decline phase

23. **In which phase of growth curve the number of cells being produced is equal to number of cells dying**
(A) Log phase
(B) Exponential phase
(C) Stationary phase
(D) Lag pahse

24. Antibiotics and other secondary metabolites are produced during which phase of growth

(A) Lag phase
(B) Log phase
(C) Decline phase
(D) Stationary phase

25. The growth phase in which the microbial population starts to decline is

(A) Lag phase
(B) Log phase
(C) Stationary phase
(D) Death phase

26. The time it takes a population to double in size is known as

(A) Population time
(B) Dividing time
(C) Generation time
(D) Binary time

27. The number of generations per unit time is known as

(A) Mean growth rate constant
(B) Generation time
(C) Mean generation time
(D) Mean growth time

28. Mean growth rate constant is expressed as

(A) Growth per hour
(B) Generations per hour
(C) Population per hour
(D) All

29. Mean generation time is

(A) Multiplication of the mean growth rate constant
(B) Reciprocal of generation time
(C) Reciprocal of the mean growth rate constant
(D) Multiplication of generation time

30. Mean generation time is denoted as

(A) k (B) t
(C) g (D) gt

31. Mean growth rate constant is denoted as

(A) kc (B) gc
(C) g (D) k

32. Generation time of *E. coli* is

(A) 20 sec (B) 20 min
(C) 30 sec (D) 30 min

33. In the culture vessel, when the nutrients are supplied at constant rate and waste products are being removed is known as

(A) Continuous culture
(B) Batch culture
(C) Fed batch culture
(D) Diauxic growth

34. The bacterial cultures can be maintained in exponential phase for longer period of time using

(A) Batch culture
(B) Fed batch culture
(C) Continuous culture
(D) Synchronous culture

35. When one of the essential nutrient (e.g. amino acid) in culture medium is limiting it is termed as

(A) Chemostat
(B) Turbidostat
(C) Diauxicstat
(D) Balanced growth

36. The system in which photocell decides the rate of addition of nutrient is known as

(A) Chemostate
(B) Diauxicstat
(C) Synchronousstat
(D) Turbidostat

37. A biphasic growth pattern or response in which a microorganism, when

exposed to two nutrients, initially uses one of them for growth and then alters its metabolism to make use of the second is known as

(A) Synchronous growth
(B) Continuous growth
(C) Diauxic growth
(D) All

38. In diauxic growth curve how many lag and log phases are found

(A) 1 (B) 2
(C) 3 (D) 4

39. A system which contain cells that are physiologically identical and are in the same stage of division cycle is known as

(A) Continuous growth
(B) Synchronous growth
(C) Diauxic growth
(D) Balanced growth

40. During which phase microorganisms are growing and dividing at maximum rate

(A) Exponential phase
(B) Stationary phase
(C) Lag phase
(D) Decline phase

41. In which phase of growth curve number of cells dividing is greater than number of cells dying

(A) Log phase
(B) Decline phase
(C) Lag phase
(D) Stationary phase

42. In which phase of growth curve, number of cells dividing is equal to number of cells dying

(A) Log phase
(B) Exponential phase
(C) Lag phase
(D) Stationary phase

43. Which of the following is not the characteristic of a growth curve?

(A) Shows development of microbial population under relatively stable environmental conditions
(B) Plotted with logarithmic numbers
(C) Graphs numbers of microbes versus time
(D) Each growth curve consists of four distinct phases

44. Nutrient content and biological structures are considered as

(A) Implicit factor for microbial growth
(B) Intrinsic factor for microbial growth
(C) Processing factor
(D) Extrinsic factors for microbial growth

45. Which of the following is used to grow bacterial cultures continuously?

(A) Chemostat
(B) Coulter Counter
(C) Hemostat
(D) Petroff-Hausser chamber

46. Chemostat operates best at .... dilution rate.

(A) High
(B) Low
(C) Moderate
(D) Medium

47. The total biomass of an organism will be determined by the nutrient present in the lowest concentration relative to the organism's requirements is known as

(A) Liebig's law of the minimum
(B) Shelford's law of tolerance
(C) Quorum sensing
(D) Heisenberg's principle of uncertainty

48. The growth is normally expressed as .... in turbidimetric measurement

(A) Cells/ml
(B) cfu/ml
(C) Optical density
(D) mg N_2 /ml

49. Which of the following procedure uses a photocell to measure absorbance of a culture to regulate the flow of culture media?

(A) Coulter Counter
(B) Hemostat
(C) Petroff-Hausser chamber
(D) Turbidostat

50. Starvation proteins are produced by a culture during which of the following segments of the growth curve?

(A) Lag phase
(B) Exponential phase
(C) Stationary phase
(D) Death phase

51. During exponential growth, the growth rate is

(A) Number of generations per unit time
(B) Reciprocal of generation time
(C) Number of generations
(D) Equivalent to generation time

52. The average size of cells during the exponential phase of growth is

(A) Greater than the lag phase
(B) Lesser than the lag phase
(C) Equal to the lag phase
(D) None of these

53. In the exponential phase, the cells and cell mass

(A) First increases then decreases
(B) Decreases
(C) Are constant
(D) Double at a constant rate

54. Postgates assay technique is useful in determining if a cell is viable even though it is

(A) Incapable of cell division
(B) Too small to see
(C) An obligate aerobe
(D) A thermophile

55. The period between inoculation of bacteria in a culture medium and beginning of multiplication is known as

(A) Stationary phase
(B) Log phase
(C) Lag phase
(D) Decline phase

56. Fragmentation occurs in

(A) *Streptomyces* species
(B) *Nocardia* species
(C) *Bacillus subtilis*
(D) *Saccharomyces cerevisiae*

57. Mean generation time is .... of mean growth rate constant

(A) Reciprocal
(B) Equivalent
(C) Multiplication
(D) Summation

58. Turbidostat operates the best at .... dilution rate

(A) High (B) Low
(C) Moderate (D) Medium

59. Dilution rate during chemostat is expressed as

(A) F=D/V (B) D=V/f
(C) D = f/V (D) F=V/D

61. The number of bacteria in aquatic samples is frequently determined by

(A) Haemocytometer
(B) Petroff-Hausser counting chambers
(C) Flow cytometer
(D) Membrane filter technique

61. Which of the following are electronic microbial cell counter

(A) Haemocytometer and Petroff-Hausser counting chambers

(B) Coulter Counter and flow cytometer

(C) Haemocytometer and Coulter Counter

(D) Haemocytometer and flow cytometer

62. Bacteria shows diauxic growth when the number of carbon sources are

(A) 1 (B) 2

(C) 3 (D) 4

63. Which of the following method can be used for viable cell count?

(A) Spectrophotometry

(B) Plate count

(C) Direct microscopy

(D) All

64. In which phase of growth curve number of cells dividing is less than number of cells dying

(A) Exponential phase

(B) Decline phase

(C) Lag phase

(D) Stationary phase

| Answer Key |

Q. No.	Ans	Q. No.	Ans	Q. No.	Ans	Q. No.	Ans	Q. No.	Ans	Q. No.	Ans
1.	A	2.	C	3.	D	4.	A	5.	C	6.	B
7.	C	8.	A	9.	C	10.	D	11.	B	12.	B
13.	D	14.	C	15.	B	16.	A	17.	D	18.	B
19.	C	20.	C	21.	A	22.	C	23.	C	24.	D
25.	D	26.	C	27.	A	28.	B	29.	C	30.	C
31.	D	32.	B	33.	A	34.	C	35.	A	36.	D
37.	C	38.	B	39.	D	40.	A	41.	D	42.	D
43.	D	44.	B	45.	A	46.	B	47.	A	48.	C
49.	D	50.	C	51.	B	52.	B	53.	D	54.	A
55.	C	56.	B	57.	A	58.	A	59.	C	60.	B
61.	B	62.	B	63.	B	64.	B				

◈

Chapter **10**

MICROBIAL NUTRITION

1. **Which macroelements or macronutrients are required for microorganisms?**
   (A) Calcium, Oxygen, Magnesium
   (B) Potassium, Calcium, Iron
   (C) Potassium, Calcium, Oxygen
   (D) Calcium, Magnesium, Iron

2. **Which macroelements in cell exist as cation**
   (A) Calcium, Oxygen, Magnesium
   (B) Potassium, Calcium, Iron
   (C) Potassium, Calcium, Oxygen
   (D) Calcium, Magnesium, Iron

3. **Which macroelement is required for activity of enzymes**
   (A) Manganese (B) Potassium
   (C) Sulphur (D) Calcium

4. **Which macroelement is involved in heat resistance of bacterial endospore**
   (A) Calcium (B) Phosphorous
   (C) Sulphur (D) Magnesium

5. **Which macroelement serves as a cofactor for many enzymes, complexes with ATP, and stabilizes ribosomes and cell membranes**
   (A) Calcium (B) Iron
   (C) Magnesium (D) Potassium

6. **Which macroelement is a part of cytochromes and a cofactor for enzymes and electron-carrying proteins**
   (A) Calcium (B) Iron
   (C) Magnesium (D) Carbon

7. **Which of the following are micronutrients or trace elements required by microbial cell**
   (A) Manganese, Cobalt, Nickel
   (B) Calcium, Iron, Magnesium
   (C) Manganese, Nickel, Magnesium
   (D) Calcium, Phosphate, Potassium

8. **Micronutrients are mainly involved in**
   (A) Structure of cofactor
   (B) Catalysis of reactions
   (C) Maintenance of protein structure
   (D) All

9. **Which micronutrient aids many enzymes catalyzing the transfer of phosphate groups**
   (A) Manganese (B) Zinc
   (C) Copper (D) Nickel

10. **Which micronutrientis required for nitrogen fixation**
    (A) Manganese (B) Molybdenum
    (C) Iron (D) Copper

11. **Which micronutrient is a part of Vitamin B12**
    (A) Copper (B) Molybdenum
    (C) Cobalt (D) Nickel

12. **Diatoms require which element for construction of their cell wall**
    (A) Carbon (B) Phosphorous
    (C) Potassium (D) Silica

13. **Which component has higher energy as compared to carbohydrate**
    (A) Protein (B) Amino acids
    (C) Lipid (D) Nucleic acid

14. Which element is the backbone of organic matter of cell

(A) Sulphur (B) Carbon

(C) Phosphorous (D) Iron

15. The organism which utilize CO_2 as sole carbon source are known as

(A) Chemotrophs (B) Lithotrophs

(C) Autotrophs (D) Heterotrophs

16. The organism which rely on preformed organic compounds for their carbon source are known as

(A) Organotrophs (B) Lithotrophs

(C) Autotrophs (D) Heterotrophs

17. The organism which use reduced organic compounds as carbon are known as

(A) Autotrophs (B) Heterotroph

(C) Organotrophs (D) Chemotrophs

18. Which organism can utilize over 100 compounds as carbon source and that's why useful in bioremediation of polluted environment

(A) *Pseudomonas aeruginosa*

(B) *Pseudomonas fluorescence*

(C) *Burkholderia cepacia*

(D) *Bacillus megaterium*

19. Which organisms utilize C_1 compounds such as methane methanol, formaldehyde etc. as carbon source

(A) Methylotrophs (B) Methanotrophs

(C) Carbonifers (D) Carbonisome

20. The organisms which use light as energy source are known as

(A) Lithotrophs (B) Phototrophs

(C) Chemotrophs (D) Autotrophs

21. The organisms which obtain energy from oxidation of chemical organic or inorganic compounds are known as

(A) Phototrophs (B) Lithotrophs

(C) Chemotroph (D) Organotrophs

22. The organism which use reduced inorganic compounds as electron source are known as

(A) Lithotroph

(B) Organotrophs

(C) Lithotrophs

(D) Autotrophs

23. The organisms which uses organic molecules as energy source are known as

(A) Autotrophs

(B) Phototrophs

(C) Organotrophs

(D) All

24. The organisms which use light energy and have CO_2 as their carbon source are known as

(A) Photoheterotrophs

(B) Chemoautotrophs

(C) Chemoorganotrophs

(D) Photoautotrophs

25. Which of the following organism is photolithotrophic autotrophs

(A) Cyanobacteria

(B) Algae

(C) Purple and green sulphur bacteria

(D) All

26. The organisms which use organic compounds as sources of energy, hydrogen, electrons, and carbon Chemoorganotrophic heterotrophs

(A) Chemoautotrophs

(B) Chemoheterotrophs

(C) Chemoorganotrophs

(D) Chemolithotrophs

27. Which of the following organisms exhibits chemoheterotrophic way of life

(A) Protozoa

(B) Fungi

(C) Non photosynthetic bacteria

(D) All

28. **The organisms which uses light as energy source and organic molecule as electron donor and carbon source are known as**
   (A) Photolithoheterotrophs
   (B) Photoorganoheterotrophs
   (C) Photoorganoautotrophs
   (D) Photolithoorganotrophs

29. **Which of the following organisms are photoorganoheterotrophs**
   (A) Purple and green non sulphur bacteria
   (B) Cyanobacteria
   (C) Algae
   (D) Iron bacteria

30. **The organisms which oxidizes reduced inorganic compounds such as iron, nitrogen, or sulfur molecules to derive both energy and electrons for biosynthesis and carbon dioxide is the carbon source are known as**
   (A) Chemoorganotroph
   (B) Chemoheterotroph
   (C) Chemolithoautotrophs
   (D) Chemoorganoheterotroph

31. **Which of the following organisms are chmolithoautotrophs**
   (A) Sulphur oxidizing bacteria
   (B) Nitrifying bacteria
   (C) Hydrogen bacteria
   (D) All

32. **The organisms which combine chemolithoautotrophic and hetero-trophic metabolic processes are known as**
   (A) Mixotrophic bacteria
   (B) Chemolithoautoheterotroph
   (C) Auxotrophs
   (D) Chemoorganotroph

33. **Which of the following organisms are mixotrophic**
   (A) Green sulphur bacteria
   (B) Purple non sulphur bacteria
   (C) Purple sulphur bacteria
   (D) Green non sulphur bacteria

34. **The process in which molecules move from a region of higher concentration to one of lower concentration because of random thermal agitation is known as**
   (A) Active diffusion
   (B) Passive diffusion
   (C) Facilitated diffusion
   (D) Assisted diffusion

35. **Which substance moves by passive diffusion**
   (A) Glycerol
   (B) Water
   (C) Gases like O_2 and CO_2
   (D) All

36. **The movement of molecules across the plasma membrane or selectively permeable barrier is dependent on**
   (A) Concentration gradient of nutrient
   (B) Membrane porosity
   (C) Membrane permeability
   (D) Cell waal ccomposition

37. **The organisms which requires temperature above 80°C are known as**
   (A) Mesophile
   (B) Thermophile
   (C) Hyperthermophile
   (D) Psychrophile

38. **When the passive diffusion is increased by using cell membrane proteins such as permeases than that phenomenon is known as**
   (A) Active diffusion
   (B) Group diffusion
   (C) Facilitated diffusion
   (D) Osmosis

39. Facilitated diffusion is more frequently observed in
(A) Prokaryotic cells
(B) Eukaryotic cells
(C) Archaebacteria
(D) Eubacteria

40. The molecules which are mainly transported through facilitated diffusion are
(A) Lipids (B) Vitamins
(C) Amino acids (D) Gases

41. Energy dependent transport processes are
(A) Active transport and group translocation
(B) Active and passive transport
(C) Passive transport and facilitated diffusion
(D) Active transport and facilitated diffusion

42. The transport of solute molecules to higher concentrations, or against a concentration gradient, with the use of metabolic energy input is known as
(A) Active transport
(B) Passive transport
(C) Facilitated diffusion
(D) Group translocation

43. ATP-binding cassette transporters (ABC transporters) are mainly involved in
(A) Facilitated diffusion
(B) Passive transport
(C) Group translocation
(D) Active transport

44. Proton gradient generated during electron transport is mainly involved in
(A) Passive transport
(B) Group translocation
(C) Active transport
(D) Facilitated diffusion

45. In the active transport, the molecules being transported remains
(A) Unaltered
(B) Modified
(C) Changed
(D) Altered

46. The transport of molecules along with proton during its inward movement is known as
(A) Antiport
(B) Symport
(C) Synport
(D) Mulpiport

47. The mechanism in which sodium ions move outside the cell in response to movement of proton inside the cell is known as
(A) Symport
(B) Synport
(C) Group translocation
(D) Antiport

48. The opposite movement of proton and sodium ions creates sodium gradient
(A) Antiport
(B) Symport
(C) Group translocation
(D) Passive transport

49. Process in which a molecule is transported into the cell while being chemically altered
(A) Facilitated diffusion
(B) Group translocation
(C) Active transport
(D) Passive transport

50. Phosphor-enol-pyruvate: sugar phosphotransferase system (PTS) for transport of carbohydrate is an example of
(A) Active transport system
(B) Passive transport system
(C) Facilitated transport system
(D) Group translocation system

51. Low molecular weight iron chealating organic compounds which aids in transport of iron in bacterial cell are known as

(A) Siderophores
(B) Iron oxidases
(C) Iron reductases
(D) Iron modulator

52. The media which contains all the ingredients of known chemical composition are known as

(A) Synthetic media
(B) Semi-Synthetic media
(C) Non-synthetic media
(D) Complex media

53. The media that contain some ingredients (e.g. peptone, yeast extract, meat extract, soybean floor etc.) of unknown chemical composition are known as

(A) Synthetic media
(B) Semi-Synthetic media
(C) Non-synthetic media
(D) Complex media

54. Agar is chemically

(A) Polysachharide (B) Lipid
(C) Protein (D) Nucleic acid

55. Agar is made up of

(A) D-galactose, 3,6-anhydro-L-galactose and D-glucuronic acid
(B) L-glucose, 3,6-anhydro-L-galactose and D-gluconic acid
(C) L-glucose, 3,4-anhydro-D-galactose and L-gluconic acid
(D) D-galactose, 3,5-hydro-D-galactose and L-glucuronic acid

56. To grow autotrophic bacteria which solidifying agent is used

(A) Agar (B) Silica
(C) Gelatine (D) Agarose

57. The media that supports growth of all type of microorganisms are known as

(A) Enrichment media
(B) Common media
(C) General purpose media
(D) Specialized media

58. Tryptic soy agar, nutrient agar are examples of

(A) General purpose media
(B) Enrichment media
(C) Specialized media
(D) Common media

59. The media in which special nutrients are added to support growth of fastidious heterotrophs are known as

(A) Simple media
(B) Enrichment media
(C) General purpose media
(D) Selective media

60. Media containing nutrients that enhance the growth and predominance of a particular type of bacterium and do not enhance other types of organisms that may be present are known as

(A) Enriched media
(B) Differential media
(C) Selective media
(D) Complex media

61. Media that distinguish between different groups of bacteria and even permit tentative identification of microorganisms based on their biological characteristics are known as

(A) Selective media
(B) Enrichment media
(C) Differential media
(D) Enriched media

62. Which of the following is selective as well as differential media

(A) Nutrient agar
(B) Mackonkey's agar
(C) Potato dextrose agar
(D) NFB media

63. **An organism has an optimal growth rate when the hydrogen ion concentration is very high. This organism is**
(A) Osmotolerant (B) Acidophile
(C) Neutrophile
(D) Aerotolerant anaerobe

64. **Which of the following has more tolerance for acidic pH (lower pH)?**
(A) Yeast and moulds
(B) Bacteria (C) *E. coli*
(D) Actinomycetes

65. **Which of the following organisms typically get their carbon for biosynthesis from carbon dioxide?**
(A) Glucose-fermenting bacteria (fermentation)
(B) Anaerobic, glucose-respiring bacteria (anaerobic respiration)
(C) Aerobic, glucose-respiring bacteria (aerobic respiration)
(D) Ammonia-oxidizing bacteria (chemolithotrophic bacteria)

66. **A culture broth tube was very turbid at the surface but clear throughout the rest of the tube indicating that the**
(A) Organisms are aerobes
(B) Organisms should be grown in an anaerobic chamber
(C) Organisms cannot produce superoxide dismutase and/or catalase
(D) Organisms cannot tolerate oxygen

67. **Bacteria of genus Nitrosomonas use .... as their electron source.**
(A) Ammonia (B) H_2S
(C) Succinate (D) Light

68. **The organism which grows best above 45°C are called**
(A) Psychrophilic (B) Mesosphilic
(C) Thermophilic (D) Alkaliphile

69. **An organism that expends energy to grow in a habitat with a low water activity in order to maintain internal solute concentrations to retain water is**
(A) Osmotolerant
(B) Acidophile
(C) Aerotolerant anaerobe
(D) Alkalophile

70. **The term facultative anaerobe refers to an organism that**
(A) Doesn't use oxygen but tolerates it
(B) Killed by oxygen
(C) Uses oxygen when present or grows without oxygen when absent
(D) Requires less oxygen than present in air

71. **The microorganisms that grow best in a low-oxygen environment is called**
(A) Aerobe (B) Anaerobe
(C) Facultative (D) Microaerophile

72. **The organisms which can grow at 0°C temperature are known as**
(A) Psychrophile
(B) Mesophile
(C) Thermophile
(D) Hyperthermophile

73. **Mesosphilic bacteria grows at temperature range**
(A) 20-30°C (B) 25-40°C
(C) >40°C (D) >55°C

74. **The term aerotolerant anaerobe refers to an organism that**
(A) Doesn't use oxygen but can grow in the presence of oxygen
(B) Killed by oxygen
(C) Requires less oxygen than is present in air
(D) Prefers to grow without oxygen

75. **Which is the differential agent in MacConkey agar ?**
(A) Bile salt
(B) Lactose and neutral red
(C) Glucose and Congo red
(D) Bile salt and Glucose

| Answer Key |

Q. No.	Ans	Q. No.	Ans	Q. No.	Ans	Q. No.	Ans	Q. No.	Ans	Q. No.	Ans
1.	A	2.	B	3.	B	4.	A	5.	C	6.	B
7.	A	8.	D	9.	A	10.	B	11.	C	12.	D
13.	C	14.	B	15.	C	16.	D	17.	B	18.	C
19.	A	20.	B	21.	C	22.	A	23.	C	24.	D
25.	D	26.	B	27.	D	28.	B	29.	A	30.	C
31.	D	32.	A	33.	B	34.	B	35.	D	36.	A
37.	C	38.	C	39.	B	40.	C	41.	A	42.	A
43.	D	44.	C	45.	A	46.	B	47.	D	48.	A
49.	B	50.	D	51.	A	52.	A	53.	B	54.	A
55.	A	56.	B	57.	C	58.	A	59.	B	60.	C
61.	C	62.	B	63.	B	64.	A	65.	D	66.	A
67.	A	68.	C	69.	A	70.	C	71.	D	72.	A
73.	B	74.	A	75.	B						

Chapter 11

MICROBIAL METABOLISM AND PHOTOSYNTHESIS

1. **The process in which larger and more complex molecules are broken down into smaller, simpler molecules with the release of energy is known as**
   (A) Metabolism (B) Anabolism
   (C) Catabolism (D) None

2. **The process in which the synthesis of complex molecules from simpler ones with the input of energy is carried out is known as**
   (A) Anabolism (B) Catabolism
   (C) Metabolism (D) None

3. **The energy yielding process in metabolism is**
   (A) Anabolism (B) Catabolism
   (C) Metabolism (D) All

4. **The energy requiring process in metabolism is**
   (A) Catabolism (B) Anabolism
   (C) Metabolism (D) All

5. **The process in which organic molecule serves as electron donor and acceptor is known as**
   (A) Fermentation (B) Respiration
   (C) Both (D) None

6. **The electron acceptor molecule in fermentation is**
   (A) Oxaloacetate (B) Carbon diaoxide (C) Pyruvate
   (D) Oxygen

7. **Energy yielding process in which electron acceptors are inorganic molecules either oxygen or nitrate, sulphate, ferric ion etc. is known as**
   (A) Respiration (B) Fermentation
   (C) Both (D) None

8. **The respiration in which final electron acceptor molecule is oxygen is known as**
   (A) Anaerobic respiration
   (B) Aerobic respiration
   (C) Fermentation
   (D) None

9. **The respiration in which final electron acceptor molecule is inorganic (e.g., NO_3, SO_4, CO_2, Fe_3, SeO_4 or fumarate etc) is known as**
   (A) Fermentation
   (B) Aerobic respiration
   (C) Anaerobic respiration
   (D) None

10. **Among fermentation and respiration which process yields higher amount of energy?**
    (A) Respiration (B) Fermentation
    (C) Both (D) None

11. **ATP are formed in respiration and fermentation through use of**
    (A) ATP transport chain
    (B) Nutrient transport chain
    (C) Electron transport chain
    (D) All

12. **Electrons for operating electron transport chain (ETC) are obtained from**
    (A) Inorganic molecules
    (B) Organic molecules
    (C) Neutral compounds
    (D) None

13. The metabolic pathway which functions both catabolically and anabolically is known as
(A) Amphibolic pathway
(B) Bidirectional pathway
(C) Multidirectional pathway
(D) All

14. Which catabolic pathways are amphibolic
(A) Glycolysis (B) TCA cycle
(C) Both (D) None

15. Which of the following pathways carry out catabolism of glucose molecules Glycolysis
(A) Entner-Doudoroff pathway
(B) Pentose phosphate pathway
(C) Both
(D) None

16. Glycolytic pathway is also known as
(A) Embden-Meyerhof pathway
(B) Entner-Doudoroff pathway
(C) Pentose phosphate pathway
(D) Calvin pathway

17. End product of glycolytic pathway is
(A) Pyruvate (B) Acetyl co A
(C) Citric acid (D) Carbon diaoxide

18. Glycolysis occurs under which conditions
(A) Aerobic (B) Anaerobic
(C) Both (D) None

19. In which part of cell glycolysis occurs
(A) Mitochondria
(B) Cytoplasmic matrix
(C) Plasma membrane
(D) Cell wall

20. When the high energy phosphate is donated to the ADP to produce ATP the process is known as
(A) Photophosphorylation
(B) Oxidative phosphorylation
(C) Substrate level phosphorylation
(D) Product level phosphorylation

21. In glycolysis, how many molecules of ATP are formed by substrate level phosphorylation
(A) One (B) Two
(C) Three (D) Four

22. In glycolysis during conversion of 1,3 bis phosphoglycerate to 3 phosphoglycerate which process occurs to generate ATP
(A) Oxidative phosphorylation
(B) Substrate level phosphorylation
(C) Either A or B
(D) Both

23. In glycolysis pathway how many ATP molecules are consumed or utilized

Or

Oxidation of one glucose molecule to pyruvate consumes how many ATPs
(A) 2 (B) 3
(C) 4 (D) 8

24. In glycolytic pathway how many NADH molecules are produced
(A) 1 (B) 2
(C) 3 (D) 4

25. Oxidation of one glucose molecule produce how many pyruvate molecules
(A) 1 (B) 2
(C) 3 (D) 4

26. During oxidation of glucose to pyruvate how many ATPs are generated?
(A) 4 (B) 6
(C) 8 (D) 10

27. Oxidation of one NADH molecule through ETC yields how many ATPs
(A) 1.5 (B) 3
(C) 4 (D) 2

28. Pentose phosphate pathway is also known as

(A) HMP pathway (B) EMP pathway
(C) ED pathway (D) EMD pathway

29. The catabolic pathway which also involves the biosynthesis is

(A) Pentose phosphate
(B) PP pathway
(C) HMP pathway
(D) All

30. Entner Doudoroff pathway is also known as

(A) ED pathway
(B) KDPG pathway
(C) Both (D) None

31. The key intermediate of ED pathway is

(A) 2-keto-3-deoxy-6-Phosphogluconate
(B) 3-keto-2-deoxy-6-Phosphogluconate
(C) 6-keto-3-deoxy-2-Phosphogluconate
(D) None

32. In ED pathway how many molecules of ATP, NADPH and NADH

(A) 1,1,1 (B) 1,2,1
(C) 2,1,1 (D) 1,1,2

33. Which pathway begins with oxidation of glucose 6 phosphate

(A) PP pathway (B) ED pathway
(C) Both (D) None

34. The terminal electron acceptor or final electron acceptor in fermentation is

(A) Pyruvate (B) Oxygen
(C) Carbon dioxide (D) Nitrate

35. In fermentation the oxidation of substrate is

(A) Full (B) Partial
(C) Either A or B (D) None

36. In fermentation the ATP is formed by

(A) Oxidative phosphorylation
(B) Reductive phosphorylation
(C) Substrate level phosphorylation
(D) Product level phosphorylation

37. In fermentation which process do not occurs

(A) Oxidative phosphorylation
(B) Substrate level phosphorylation
(C) Reductive phosphorylation
(D) Product level phosphorylation

38. In fermentation oxidative phosphorylation do not occurs due to

(A) Absence of oxygen
(B) Absence of pyruvate
(C) Absence of carbon dioxide
(D) Absence of nitrogen

39. In fermentation which pathway do not operate to produce ATP

(A) Oxidative phosphorylation
(B) Glycolysis
(C) Substrate level phospholrylation
(D) Electron transport chain (ETC)

40. The pyruvate molecule generated in glycolysis is converted in to carbon dioxide through

(A) ED pathway
(B) EMP pathway
(C) TCA cycle
(D) EMD pathway

41. The substrate for TCA cycle is

(A) Pyruvate
(B) Acetyl Co A
(C) Citric acid
(D) Glucose

42. Conversion of pyruvate to Acetyl Co A is carried out by

(A) Pyruvate oxidase
(B) Acetyl Co A synthase
(C) Pyruvate dehydrogenase complex
(D) Citrate dehydrogenase complex

43. **TCA cycle operates under**
(A) Anerobic condition
(B) Aerobic condition
(C) Acidic condition
(D) Alkaline condition

44. **The final product of TCA cycle is**
(A) Oxalate (B) Oxaloacetate
(C) Acetate (D) Citrate

45. **ETC is located in which part of the bacterial cell**
(A) Cell wall
(B) Cytoplasm
(C) Mitocondria
(D) Plasma membrane

46. **ETC in eukaryotic cells are located in**
(A) Outer mitochondrial membrane
(B) Inner mitochondrial membrane
(C) Mitochondrial matrix
(D) Plasma membrane

47. **The system composed of a series of electron carriers that operate together to transfer electrons from donors, like NADH and FADH2, to acceptors, such as O_2 and produce ATP is known as**
(A) CTC (B) ECT
(C) DTC (D) ETC

48. **How many electron carrier complexes are there in ETC**
(A) 4 (B) 3
(C) 2 (D) 1

49. **Which molecules connect ETC complexes with each other**
(A) Cytochromes b and c
(B) Cytochrome Q and b
(C) Coenzyme Q and cytochrome b
(D) Coenzyme Q and cytochrome c

50. **The process by which energy from electron transport is used to make ATP is called**
(A) Oxidative phosphorylation
(B) Substrate level phosphorylation
(C) Oxidative photophosphorylation
(D) photophosphorylation

51. **When electron molecule passes from NADH to oxygen how many ATPs are generated**

Or

NADH yields how many ATP
(A) 2 (B) 1
(C) 3 (D) 4

52. **Yield of ATP from NADH and FADH molecule depends on**
(A) Phosphorous: Oxygen ratio (P/O ratio)
(B) Carbon: Nitrogen ratio (C/N ratio)
(C) Oxygen: Sulphur ration (O/S ratio)
(D) Phosphorous: Carbon ration (P/C ratio)

53. **When electron molecule passes from FADH2 to oxygen how many ATPs are generated**

Or

FADH2 yields how many ATP
(A) 1 (B) 2
(C) 3 (D) 4

54. **The mechanism of oxidative phosphorylation is best explained by**
(A) Chemiluminous hypothesis
(B) Chemiosmotic hypothesis
(C) Osmotic hypothesis
(D) Osmochemical hypothesis

55. **In oxidative phosphorylation electron moves outside and ATP moves inward with the help of**
(A) ATP motive force (AMF)
(B) Proton motive force (PMF)
(C) Oxygen motive force (OMF)
(D) Electron motive force (EMF)

56. **ATP synthesis takes place by using enzyme**
(A) ATPase (B) ATP synthase
(C) F0F1ATPase (D) All of above

57. ATP synthesis takes place on which part of ATPase

(A) α subunit (B) β subunit

(C) γ subunit (D) δ subunit

58. 'γ subunit' contain how many active sites

(A) 1 (B) 2

(C) 3 (D) 4

59. F0 subunit is connected with F1 subunit through

(A) ε subunit and γ subunit

(B) α subunit and β subunit

(C) α subunit and ε subunits

(D) γ subunit and β subunit

60. F1 subunit lies in

(A) Cytoplasm (B) Membrane

(C) Matrix (D) Ribosome

61. Which chemicals inhibit electron transport in ETC directly and thereby inhibit ATP synthesis

(A) Antibiotic piericidin and antimycin A

(B) Cyanide

(C) Azide

(D) All of above

62. Antimycin A inhibits transport of electron between

(A) Cytochromes b and c

(B) Cytochrome Q and b

(C) Coenzyme Q and cytochrome b

(D) Coenzyme Q and cytochrome c

63. Cyanide and azide stop the transfer of electrons between

(A) Coenzyme Q and cytochrome c

(B) Cytochrome a and O_2

(C) Cytochrome c and O_2

(D) Coenzyme Q and O_2

64. Which chemicals are structural analogue of O_2

(A) Cyanide and Azide

(B) Carbon monosulphide and Salacinol

(C) Cyanide and Salacinol

(D) Azide and Salacinol

65. Chemicals which inhibits the electron transport chain without inhibiting transport of electrons are known as

(A) Coupler (B) Uncouplers

(C) Couplings (D) Co-operators

66. Which chemicals act as uncoupler in ETC

(A) Dinitrophenol and Valinomycin

(B) Cyanide and Azide

(C) Antibiotic piericidin and antimycin A

(D) Salacinol and cyanide

67. Many microorganisms, when moved from anaerobic to aerobic conditions, will drastically reduce their rate of sugar catabolism and switch to aerobic respiration, this phenomenon is known as

(A) Leibig's effect (B) Pasteur effect

(C) Reverse effect (D) None of above

68. When the electron transport chain operates with exogenous electron acceptor the process is known as

(A) Aerobic respiration

(B) Oxygenic respiration

(C) Anaerobic respiration

(D) Fermentation

69. The major electron acceptors in anaerobic respiration are

(A) Nitrate, Sulfate and CO_2

(B) O_2, SO and NO

(C) CO_2, H_2O and O_2

(D) CO, NO, O

70. Process in which light energy is trapped and converted to chemical energy is called

(A) Respiration

(B) Photophosphorylation

(C) Electron transport

(D) Photosynthesis

71. Photosynthesis can be divide in how many parts

(A) 2 (B) 3
(C) 4 (D) 5

72. The process in which light energy is trapped and converted to chemical energy

(A) Dark reaction
(B) Light reaction
(C) Green raction
(D) Pasteur reaction

73. The process in which chemical energy is used to reduce or fix CO_2 and synthesize cell constituents is known as

(A) Light reaction
(B) White reaction
(C) Dark reaction
(D) Black reaction

74. Which pigments are accessory in photosynthetic bacteria

(A) Carotenoid
(B) Phycobiliproteins
(C) Both
(D) None

75. Accessory pigment can absorb which colored light

(A) Blue green to yellow
(B) Green to red
(C) Blue to green
(D) Blue to yellow

76. In eukaryotes and cyanobacteria how many light trapping reaction centres or photosystems are there

(A) Photosystem I and Photosystem II
(B) Photosystem II and Photosystem III
(C) Photosystem I and Photosystem III
(D) Photosystem I and Photosystem IV

77. PS I is also known as

(A) P700 (B) P680
(C) P860 (D) P660

78. The photosynthetic process in which electrons moves though series of electron carriers and come back to same photosystem the process is known as

Or

The process in which only one photosystem participates

(A) Cyclic photophosphorylation
(B) Circular photophosphorylation
(C) Cyclic photophosphorylation
(D) Acircular photophosphorylation

79. The photosynthetic process in which electrons moves though series of electron carriers in to both the photosystems, the process is known as

(A) Cyclic photophosphorylation
(B) Non cyclic photophosphorylation
(C) Circular photophosphorylation
(D) Non circular photophosphorylation

80. In non-cyclic photophosphorylation light reaction generates how many ATP and NADH molecules

(A) 2,2 (B) 2,3
(C) 3,2 (D) 2,4

81. In bacteria which type of photophosphorylation takes place

(A) Non cyclic (B) Cyclic
(C) Both (D) None

82. In green and purple sulphur bacteria (photosynthetic bacteria) which photosystem is present

(A) PS I (B) PS II
(C) PS III (D) PS IV

83. Bacterial photosynthesis is

(A) Oxygenic (B) Anoxygenic
(C) Both (D) None

84. Eukaryotic and cyanobacterial photosynthesis is

(A) Oxygenic (B) Anoxygenic
(C) Both (D) None

85. **The process of protein breakdown is**
(A) Anabolic process
(B) Catabolic process
(C) Metabolic process
(D) All of above

86. **Protease enzyme breakdown which bond that holds two amino acids together and thereby form protein**
(A) Phoshodiester bond
(B) Hydrogen bond
(C) Peptide bond
(D) Covalent bond

87. **The process of removal of the amino group from an amino acid is known as**
(A) Transamination
(B) Amine transformation
(C) Ammonification
(D) Deamination

88. **The process in which amino group is transferred from an amino acid to an -keto acid acceptor is known as**
(A) Deamination
(B) Transamination
(C) Ammonification
(D) None of above

89. **During the carboxylation phase of the Calvin cycle, CO_2 combines with**
(A) Ribulose 1,5 - bisphosphate
(B) Phosphoglyceraldehyde
(C) Pyruvic acid
(D) Oxaloacetic acid

90. **Which of the following groups contain(s) many unique coenzymes, such as coenzyme M and coenzyme F420?**
(A) Sulfate-reducing bacteria
(B) Methanotrophs (methane-oxidizing microbes)
(C) Methanogens (methane-producing microbes)
(D) Acetogens (acetigens; acetate-producing microbes)

91. **In an oxygenic photosynthesis, the green and the purple bacteria do not use which of the following one as an electron source?**
(A) H_2O (B) H_2
(C) H_2S (D) S (elemental sulphur)

92. **Assimilatory sulfate reduction involves the nucleotide .... during the incorporation of H_2S in the production of ....**
(A) ATP; methionine
(B) ATP; cytosine
(C) UTP; cytosine
(D) GTP; cytosine

93. **The chlorophyll molecules used by eucaryotes and cyanobacteria absorb radiant energy in the .... portion(s) of the visible spectrum.**
(A) Red
(B) Green
(C) Red and blue
(D) Green and ultraviolet

94. **The acquisition energy by glucose fermentation requires**
(A) Substrate-level phosphorylation
(B) Electron transport of electrons from NADH
(C) Long-chain fatty acid oxidation
(D) Enzyme formic-hydrogen lyase

95. **In establishing proton gradient for chemiosmotic ATP generation by aerobic respiration the terminal electron acceptor is**
(A) Nitrate (B) Oxygen
(C) Sulfate (D) CO_2

96. **If ΔG of a chemical reaction is positive in value and keq is less than 1 then the chemical reaction will**
(A) Proceed in reverse direction
(B) Proceed in forward direction
(C) Not take place in any of the direction
(D) Does not dependent

97. Which of the following nucleoside diphosphates is used most often in carbohydrate anabolism?
(A) Uridine diphosphate
(B) Adenosine diphosphate
(C) Guanine diphosphate
(D) Thymine diphosphate

98. In which cycle microorganisms use acetate as sole source of carbon?
(A) Pentose phosphate pathway
(B) Glycolyic pathway
(C) Glyoxylate pathway
(D) Oxaloacetate pathway

99. The role of bacteriophyll in an oxygenic photosynthesis is to
(A) Reduce ferridoxin directly
(B) Reduce NADP directly
(C) Use light energy to energize an electron
(D) Transfer electrons to an intermediate in the sulfide oxidation pathway

100. Which of the following(s) is/are the products of the light reactions of photosynthesis?
(A) ATP only
(B) NADPH only
(C) ATP and O_2 only
(D) ATP, NADPH, and O_2

101. As the electron flow through the chains, much of their free energy is conserved in the form of ATP. This process is called
(A) Oxidative phosphorylation
(B) Electromotive potential
(C) Dehydrogenations
(D) Decarboxylation

102. Most of the energy in aerobic respiration of glucose is captured by
(A) Substrate-level phosphorylation
(B) Electron transport of electrons from NADH
(C) Long-chain fatty acid oxidation
(D) Enzyme formic-hydrogen lyase

103. During the reduction phase of the Calvin cycle, phosphoglyceric acid is reduced to .... utilizing .... as the reduction source.
(A) Phosphoglyceraldehyde; $NADPH^+$, H^+
(B) Phosphoglyceraldehyde; $NADH^+$, H^+
(C) Ribulose 1,5 - bisphosphate; $NADH^+$, H^+
(D) Pyruvic acid; $NADPH^+$, H^+

104. The major route for incorporation of ammonia (NH_4^+) into organic compounds is via
(A) Reduction of pyruvate or alpha-ketoglutarate by enzymes
(B) Atmospheric nitrogen fixation
(C) Oxidation of pyruvate
(D) Reduction of pyruvate

105. The bacteriochlorophylls used by the anoxygenic bacteria have absorbance maxima located in the .... portion(s) of the spectrum.
(A) Green (B) Blue
(C) Ultraviolet (D) Infrared

106. The mechanism of passive or facilitated diffusion require
(A) Metabolic energy
(B) Concentration of solute against an electrochemical gradient
(C) Accumulation of solute against an electrochemical gradient
(D) Accumulation or concentration of solute against an electrochemical gradient

107. At the end of transamination reaction which molecules are formed
(A) Amino acid (B) Organic acid
(C) Nucleic acid (D) None

| Answer Key |

Q. No.	Ans	Q. No.	Ans	Q. No.	Ans	Q. No.	Ans	Q. No.	Ans	Q. No.	Ans
1.	C	**2.**	A	**3.**	B	**4.**	B	**5.**	A	**6.**	C
7.	A	**8.**	B	**9.**	C	**10.**	A	**11.**	C	**12.**	A
13.	A	**14.**	C	**15.**	C	**16.**	A	**17.**	A	**18.**	C
19.	B	**20.**	C	**21.**	B	**22.**	B	**23.**	A	**24.**	B
25.	B	**26.**	C	**27.**	B	**28.**	A	**29.**	D	**30.**	C
31.	A	**32.**	A	**33.**	C	**34.**	A	**35.**	B	**36.**	C
37.	A	**38.**	A	**39.**	D	**40.**	C	**41.**	B	**42.**	C
43.	B	**44.**	B	**45.**	D	**46.**	B	**47.**	D	**48.**	A
49.	D	**50.**	A	**51.**	C	**52.**	A	**53.**	B	**54.**	B
55.	B	**56.**	D	**57.**	B	**58.**	C	**59.**	A	**60.**	C
61.	D	**62.**	A	**63.**	B	**64.**	A	**65.**	B	**66.**	A
67.	B	**68.**	C	**69.**	A	**70.**	D	**71.**	A	**72.**	B
73.	C	**74.**	C	**75.**	A	**76.**	A	**77.**	A	**78.**	A
79.	B	**80.**	A	**81.**	B	**82.**	A	**83.**	B	**84.**	A
85.	B	**86.**	C	**87.**	D	**88.**	B	**89.**	A	**90.**	C
91.	A	**92.**	B	**93.**	C	**94.**	A	**95.**	B	**96.**	A
97.	A	**98.**	C	**99.**	C	**100.**	D	**101.**	A	**102.**	B
103.	A	**104.**	A	**105.**	D	**106.**	A	**107.**	B		

◈

Chapter 12

SOIL MICROBIOLOGY

1. **Which soil fungus is found associated with tree roots in hard wood forest**
   (A) *Apergillus flavus*
   (B) *Penicillium notatum*
   (C) *Armillaria bulbosa*
   (D) *Chaetomium globosum*
2. **How much portion of soil microorganism is cultured**
   (A) 30 % (B) 70 %
   (C) 90 % (D) 10 %
3. **Gram positive bacteria occurring in soil are belonging to which groups**
   (A) Coryneforms
   (B) Nocardioforms
   (C) Actinomycetes
   (D) All of above
4. **The compound produced by actinomycete that gives the characteristic earthy odour to soil is**
   (A) Gastrosine
   (B) Geosmine
   (C) Jasmonic acid
   (D) Geosmic acid
5. **Which soil microorganism respond rapidly to the addition of easily usable source of carbohydrate i.e. sugar and amino acids**
   (A) *Pseudomonas* (B) *Bacillus*
   (C) *Azotobacter* (D) *Acetobacter*
6. **Which of the following microorganism utilize soil organic matter**
   (A) *Arthrobacter* and *Actinomycetes*
   (B) *Acinetobacterium* and *Azotobacter*
   (C) *Acetobacter* and *Azotobacter*
   (D) *Acenetobacterium* and *Arthrobacter*
7. **Volatile compounds released by plant roots are**
   (A) Ethanol and isobutanol
   (B) Isoamyl alcohol and acetoin
   (C) Isobutyric acid and ethylene
   (D) All of above
8. **Low molecular weight compounds released by plant roots are**
   (A) Sugars and organic acids
   (B) Amino acids and nucleotides
   (C) Vitamins
   (D) All of above
9. **High molecular weight compounds released by plant roots are**
   (A) Polysachharides and Enzymes
   (B) Sugars and organic acids
   (C) Amino acids and nucleotides
   (D) Amino acids and vitamins
10. **The volume of soil around the root influenced by root exudates is known as**
    (A) Rhizosphere (B) Rhizoplane
    (C) Rhizosoil (D) Bulk soil
11. **Free living nitrogen fixing bacteria in soil is known as**
    (A) *Acetobacter* (B) *Azospirillum*
    (C) *Azotobacter* (D) *Rhizobium*
12. **Associative nitrogen fixing bacteria is**
    (A) *Rhizobium* (B) *Azospirillum*
    (C) *Frankia* (D) *Acetobacter*

13. Nodule forming (symbiotic) nitrogen fixing bacteria is

(A) *Rhizobium* and *Allorhizobium*
(B) *Azorhizobium* and *Bradyrhizobium*
(C) *Mesorhizobium* and *Sinorhizobium*
(D) All of above

14. Which plant genes are important for nodule establishment

(A) *Nin* (B) *Nif*
(C) *Nod* (D) *Bac*

15. Which is the signal molecule secreted by plant to stimulate *Rhizobium* for nodule formation is

(A) Flavonoid (B) Nod factors
(C) Nitrogenase (D) Alkaloids

16. Flavonoid activate which factor that stimulate symbiotic process

(A) Nod factor (B) Nif genes
(C) Nin genes (D) Nil genes

17. Inside the plant *Rhizobium* occur in the form of

(A) Bacteroid (B) Single bacteria
(C) Spore (D) Cyst

18. Which pigment in *Rhizobium* protect nitrogenase form oxygen deactivation

(A) Leghaemoglobin
(B) Haemoglobin
(C) Phycocyanin
(D) Phycobilin

19. Nitrogen fixing microorganisms possess which enzyme for nitrogen fixation

(A) Hydrogenase
(B) Nitrite reductase
(C) Nitrogenase
(D) Nitrase

20. Nitrogenase enzyme is sensitive to

(A) Carbon Dioxide
(B) Oxygen
(C) Carbon monoxide
(D) Hydrogen

21. In mature root nodule which structures are site of nitrogen fixation

(A) Symbiosome
(B) Bacteroid
(C) Leghaemoglobin
(D) Both A and B

22. The concept which defines the property of *Rhizobium* that after developed into the final differentiated bacteria, they no longer can revert back to bacteria capable of reproduction and have been lost to the "gene pool." Is known as

(A) Altruism in the rhizosphere
(B) Gene pool of rhizosphere
(C) Gene alteration of rhizosphere
(D) Alteration of exudates

23. The bacterial community, which have achieved nodulation and thus have sacrificed their own reproduction is

(A) *Frankia* (B) *Rhizobium*
(C) *Acetobacter* (D) *Azospirillum*

24. Plant protein that is involved in attachment of *Rhizobium* to root hair is

(A) Cellulose (B) Lignin
(C) Chitin (D) Lectin

25. Bacteria which form stem nodules on *Sesbania rosrata* are

(A) *Azorhizobium*
(B) *Bradyrhizobium*
(C) *Rhizobium*
(D) *Synorhizobium*

26. Mycorrhizae was first discovered by

(A) S. Windogradsky
(B) Louis Pasteur
(C) A. B. Frank
(D) Robert Koch

27. Mycorrhizae plant association is

(A) Symbiotic (B) Synergistic
(C) Parasitic (D) Predatory

28. Which type of mycorrhizae produce Hartig net
(A) Endomycorrhizae
(B) Ectomycorrhizae
(C) Ectendomycorrhizae
(D) Exomycorrhizae

29. Which of the following are the genus of Hartig net producing mycorrhizae or ectomycorrhizae
(A) Cennococcum (B) Pisolithus
(C) Amanita (D) All of above

30. The mycorrhizae which produces only sheath is known as
(A) Ectendomycorrhizae
(B) Endomycorrhizae
(C) Ectomycorrhizae
(D) Exomycorrhizae

31. The mycorrhizae which produce intracellular coils are known as
(A) Ectomycorrhizae
(B) Endomycorrhizae
(C) Ectendomycorrhizae
(D) Exomycorhhizae

32. The mycorrhizae which produces both sheath and intracellular coils are known as
(A) Ectomycorrhizae
(B) Endomycorrhizae
(C) Ectendomycorrhizae
(D) Exomycorrhizae

33. Ectomycorrhizae fungi generally belongs to which class of fungi
(A) Ascomycetes
(B) Zygomycetes
(C) Duteromycetes
(D) Basidiomycetes

34. Endomycorrhizal fungi generally belongs to which class of fungi
(A) Ascomycetes
(B) Zygomycetes
(C) Duteromycetes
(D) Basidiomycetes

35. Genus of VAM are
(A) Glomus and Gigaspora
(B) Goniudium and Glomus
(C) Gonidium and Gigaspora
(D) Globulus and Glomus

36. Mycorrhizae increase availability of which main nutrient for crops
(A) Phosphorous (B) Nitrogen
(C) Potash (D) Iron

37. Symbiotic nitrogen fixing actinomycetes associated with trees and shrubs are
(A) *Frankia* (B) *Streptomyces*
(C) *Rhizobium* (D) *Azorhizobium*

38. Frankia forms nodules on
(A) Bayberries and alders
(B) Alnus and Ceanothus
(C) Casuarina and Comptonia
(D) All

39. Stem nodulating *Rhizobium* is
(A) *Azorhizobium*
(B) *Bradyrhizobium*
(C) *Synorhizobium*
(D) *Rhizobium*

40. Tumor inducing plasmid of agrobacterium is known as
(A) Ri Plasmid (B) Tu Plasmid
(C) Ti plasmid (D) Tm Plasmid

41. Tumour inducing genes of plant are known as
(A) Oncogenes (B) Nif genes
(C) Nod genes (D) Nin genes

42. Upon infection by *Agrobacterium* plants synthesize which component that helps in proliferation of *Agrobacterium*
(A) Flavonoids (B) Tumorosine
(C) Opines (D) Opens

43. An area between decaying plant material and soil is known as
(A) Rhizosphere (B) Rhizoplane
(C) Bulk soil (D) Residuesphere

44. Degradation of which plant component requires oxygen
(A) Lignin (B) Cellulose
(C) Hemicellulose (D) All

45. Which of the following enzyme is important in degradation of lignin?
(A) Ligninase
(B) Laccase
(C) Polyphenol oxidase
(D) All of above

46. During degradation of woody plant which unusual or undesired gases are produced
(A) Chloromethane
(B) Cyanide
(C) Both A and B
(D) None

47. Anaerobic nitrogen fixing bacteria are
(A) *Clostridium* (B) *Rhizobium*
(C) *Acetobacter* (D) *Azotobacter*

48. *Azotobacter* protects its nitrogenase from oxygen by
(A) High respiration rate
(B) Microaerophilic respiration
(C) Leghaemoglobin
(D) Nodulation

49. Microaerophilic associative nitrogen fixing bacteria is
(A) *Azotobacetr* (B) *Acetobacter*
(C) *Rhizobium* (D) *Azospirillum*

50. The organisms which require 21% oxygen for their growth are
(A) *Azotobacter* (B) *Azospirillum*
(C) *Rhizobium* (D) *Acetobacter*

51. *Azospirillum* protects nitrogenase from oxygen deactivation by
(A) Leghaemoglobin
(B) Thick walled cells
(C) Microaerophilic respiration
(D) Nodulation

52. The nitrogen fixing structures of blue green algae is
(A) Akinites (B) Nodules
(C) Vesicles (D) Heterocyst

53. Heterocyst limits the rate of oxygen diffusion and thereby protect nitrogenase from oxygen inactivation due to
(A) Thick wall
(B) Leghaemoglobin
(C) High respiration rate
(D) Microaerophillic respiration

54. Nitrogen fixing water fern is
(A) Lotus (B) Azolla
(C) Anabaena (D) All of above

55. Most widely used strain of *Azolla* for nitrogen fixation is
(A) *Azolla pinnata* (B) *Azolla pinia*
(C) *Azolla azollae* (D) All of above

56. The media used to grow *Azotobacter* is
(A) Ashby's media
(B) Jenson's media
(C) Burk's media
(D) All of above

57. PSB solubilize insoluble form of phosphate such as tricalcium phosphate, alluminium phosphate etc by mechanism of
(A) Organic acid production
(B) Phosphatase production
(C) Phytase production
(D) Amino acid production

58. *Azotobacter* fixes how much nitrogen in soil
(A) 100kg/ha (B) 20 kg/ha
(C) 50 kg/ha (D) 80 kg/ha

59. *Acetobacter* can fix nitrogen up to
(A) 200 kg/ha (B) 500 kg/ha
(C) 800 kg/ha (D) 870 kg/ha

60. A heterocyst is
(A) Type of spore
(B) Terminally differentiated cell that fixes nitrogen
(C) Progenitor of cyanobacterial vegetative cells
(D) Cell that carries out oxygenic photosynthesis

61. Azolla harbors which blue green algae?
(A) *Anabaena cycadae*
(B) *Anabaena azollae*
(C) *Anabaena circinalis*
(D) *Anabaena aequalis*

62. The diagnostic enzyme for denitrification is
(A) Nitrate reductase
(B) Nitrate oxidase
(C) Nitro oxidoreductase
(D) Nitrogenase

63. Which of the following organisms play a key role in the transformation of rock to soil.
(A) Cyanobacteia
(B) Pectin decomposing bacteria
(C) Nitrifying bacteria
(D) De-nitrifying bacteria

64. For rapid decomposition by microbes, the substrate should have a C/N ratio of
(A) 10-20 (B) 20-30
(C) 30-40 (D) 60-80

65. Humus and the smallest particles of rock form the
(A) A-horizon (B) B-horizon
(C) C-horizon (D) Bedrock

66. Which of the following soil microorganism is involved in the reduction of sulfates to H_2S?
(A) *Thiobacillus thiooxidans*
(B) *Desulfotomaculum*
(C) *Rhodospirillum*
(D) *Rhodomicrobium*

67. Hue denotes
(A) Dominant spectrum
(B) Lightness or brightness
(C) Purity
(D) Intensity

68. Which organic fraction not soluble in both acid and alkali
(A) Fulvic acid (B) Humic acid
(C) Humin (D) Humus

69. Which genera of bacteria found maximum in soil?
(A) *Azotobacter* (B) *Azospirillum*
(C) *Pseudomonas* (D) *Bacillus*

70. Which of the following organic fraction is soluble in alkali and insoluble in acid
(A) Fulvic acid (B) Humic acid
(C) Humin (D) Humus

71. Which of the following organic fraction is soluble in acid and alkali both
(A) Fulvic acid (B) Humic acid
(C) Humin (D) Humus

72. Which of the following organic fraction is having greatest resistance to microbial action
(A) Fulvic acid (B) Humic acid
(C) Humin (D) Humus

73. Which of the following organic fraction is susceptible to microbial action
(A) Fulvic acid (B) Humic acid
(C) Humin (D) Humus

74. The peak microbial population was observed in soil generally after addition of residue
(A) 3-4 week (B) 2-3 week
(C) 3-4 days (D) 1-2 week

75. SOM is generally believed to contain ....% C
(A) 58 (B) 62
(C) 65 (D) 68

| Answer Key |

Q. No.	Ans	Q. No.	Ans	Q. No.	Ans	Q. No.	Ans	Q. No.	Ans	Q. No.	Ans
1.	C	2.	D	3.	D	4.	B	5.	A	6.	A
7.	D	8.	D	9.	A	10.	A	11.	C	12.	B
13.	D	14.	A	15.	A	16.	A	17.	A	18.	A
19.	C	20.	B	21.	D	22.	A	23.	B	24.	D
25.	A	26.	C	27.	A	28.	B	29.	D	30.	C
31.	B	32.	C	33.	D	34.	B	35.	A	36.	A
37.	A	38.	D	39.	A	40.	C	41.	A	42.	C
43.	D	44.	A	45.	D	46.	C	47.	A	48.	A
49.	D	50.	B	51.	C	52.	D	53.	A	54.	B
55.	A	56.	D	57.	A	58.	B	59.	A	60.	B
61.		62.	A	63.	A	64.	C	65.	C	66.	B
67.	A	68.	C	69.	D	70.	B	71.	A	72.	C
73.	A	74.	D	75.	A						

◈

Chapter 13

NITROGEN CYCLE

1. **The process of oxidation of ammonia (NH_3) to Nitrate (NO_3) is known as**
   (A) Denitrification
   (B) Ammon*if*ication
   (C) Nitrification
   (D) Nitrogen fixation
2. **First step of nitrification is conversion of ammonia to**
   (A) Nitrite (NO_2)
   (B) Nitrate (NO_3)
   (C) Dinitrogen (N_2)
   (D) Nitrous oxide (N_2O)
3. **Second step of nitrification process is conversion of nitrite to**
   (A) Nitrous oxide (N_2O)
   (B) Nitrite (NO_2)
   (C) Nitrate (NO_3)
   (D) Dinitrogen (N_2)
4. **Process of nitrification is**
   (A) Anaerobic
   (B) Microaerophilic
   (C) Aerobic
   (D) Oxygenic
5. **The bacteria involved in conversion of ammonia to nitrite is**
   (A) *Nitrosomonas and Nitrosococcus*
   (B) *Nitrobacter and Nitrococcus*
   (C) *Nitrobacter and Nitrosomonas*
   (D) *Nitrosococcus and Nitrococcus*
6. **Bacteria involved in conversion of nitrite to nitrate is**
   (A) *Nitrosomonas and Nitrosococcus*
   (B) *Nitrobacter and Nitrococcus*
   (C) *Nitrobacter and Nitrosomonas*
   (D) *Nitrosococcus and Nitrococcus*
7. **Name the bacteria which carry out oxidation of ammonia to nitrite and nitrous oxide anaerobically using nitrogen dioxide as oxidant**
   (A) *Nitrosococcus eutropha*
   (B) *Nitrosomonas eutropha*
   (C) *Nitrosobacter eutropha*
   (D) *Nitrobacter eutropha*
8. **In more acidic environment which type of nitrification occurs**
   (A) Autotrophic
   (B) Heterotrophic
   (C) Autoheterotrophic
   (D) Chemoautotroph
9. **Heterotrophic nitrification is carried out by**
   (A) Bacteria and Fungi
   (B) Fungi and Algae
   (C) Algae and Actionomycetes
   (D) Cyanobacteria and Algae
10. **In nitrification process electron donor for *Nitrosomonas* species is**
    (A) Ammonium (B) Nitrogen gas
    (C) Nitrate (D) Nitrite
11. **In nitrification process electron donor for *Nitrobacter* is**
    (A) Nitrite (B) Nitrate
    (C) Ammonia (D) Nitrogen
12. **In nitrification electron acceptor of *Nitrosomonas* is**
    (A) Ammonia (B) Nitrogen gas
    (C) Nitrate (D) Nitrite

13. In nitrification electron acceptor molecule for *Nitrobacter* is

(A) Nitrite (B) Ammonium
(C) Nitrate (D) Pyruvate

14. The process in which nitrate is reduced to nitrite or nitrogen gas is known as

(A) Assimilatory nitrate reduction
(B) Assimilatory nitrite reduction
(C) Dissimilatory nitrate reduction
(D) Dissimilatory nitrite reduction

15. The process of conversion of nitrate to nitrous oxide (N_2O) or nitrogen (N_2) is known as

(A) Nitrification
(B) Denitrification
(C) Ammonification
(D) Nitrogen fixation

16. Denitrification is carried out by which bacteria

(A) Auxotrophs (B) Heterotrophs
(C) Autotrophs (D) Mixotrophs

17. The process in which nitrate is converted to ammonia is known as

(A) Dissimilatory reduction
(B) Assimilatory reduction
(C) Dissimilatory oxidation
(D) Assimilatory oxidation

18. Conversion of nitrate to ammonia is carried out by

(A) Geobacter metallireducens
(B) *Desulfovibrio* spp.
(C) *Clostridium* spp.
(D) All of above

19. Anaerobic oxidation of NH_4 with the reduction of NO_2, to produce gaseous nitrogen is known as

(A) Nitrification
(B) Assimilatory nitrate reduction
(C) Dissimilatory nitrate reduction
(D) Anamox reaction

20. The process of anoxic ammonia oxidation is known as

(A) Anamox reaction
(B) Ammonification
(C) Nitrification
(D) Nitrogen fixation

21. The process of conversion of nitrate to gaseous nitrogen is known as

(A) Nitrification
(B) Nitrate oxidation
(C) Nitrate reduction
(D) Denitrification

22. In nitrogen cycle which process is disadvantageous process

Or

Undesired process in nitrogen cycle is

(A) Nitrification
(B) Ammonification
(C) Denitrification
(D) Nitrogen fixation

23. The process in which nitrate is used as alternative electron acceptor in energy generation is known as

(A) Assimilatory nitrate reduction
(B) Dissimilatory nitrate reduction
(C) Nitrification
(D) Denitrification

24. The process in which nitrate is reduced to oxidation level of ammonia for use as a source of nitrogen for growth is known as

(A) Dissimilatory nitrate reduction
(B) Assimilatory nitrate oxidation
(C) Dissimilatory nitrogen oxidation
(D) Assimilatory nitrate reduction

25. The process of denitrification is

(A) Obligate anaerobic
(B) Obligate aerobic
(C) Facultative aerobic
(D) Facultative anaerobic

26. The enzyme involved in denitrification process is
(A) Nitrate reductase
(B) Nitrite reductase
(C) Nitrogenase
(D) Dinitrogenase reductase

27. Assimilatory nitrate reduction takes place under
(A) Microaerophilic condition
(B) Anaerobic condition
(C) Aerobic conditions
(D) Dark condition

28. Assimilatory nitrate reduction takes place in
(A) Plants (B) Fungi
(C) Bacteria (D) All of above

29. Dissimilatory nitrate reduction takes place in
(A) Bacteria (B) Plants
(C) Fungi (D) Nematodes

30. When denitrification occurs under anaerobic condition which gas is formed
(A) Nitrogen
(B) Ammonia
(C) Nitrous oxide
(D) Nitrogen dioxide

31. When denitrification occurs under microaerobic condition which gas is formed
(A) Dinitrogen
(B) Nitrous oxide
(C) Ammonia
(D) Nitrogen dioxide

32. Under low pH which gas is formed as a result of denitrification
(A) Nitrous oxide
(B) Ammonia
(C) Dinitrogen
(D) Nitrogen dioxide

33. Rate of denitrification gets lowered down under
(A) Neutral pH
(B) Alkaline pH
(C) Acidic pH
(D) Near neutral pH

34. When the nitrate concentration is higher, production of which gas is favoured
(A) Nitrogen
(B) Nitrous oxide
(C) Ammonia
(D) Nitrogen dioxide

35. When nitrate concentration is higher and organic energy sources are there than denitrification results in the formation of which gas
(A) Nitrogen
(B) Nitrogen dioxide
(C) Nitrous oxide
(D) Ammonia

36. The process of ammonia production due to degradation of organic nitrogen compounds is known as
(A) Denitrification
(B) Nitrification
(C) Ammonification
(D) Nitrogenase

37. Under anaerobic conditions ammonification results in the formation of
(A) Nitrate ions
(B) Nitrite ions
(C) Ammonia ions
(D) Nitrogen ion

38. Under aerobic condition ammonia undergoes
(A) Ammonification
(B) Denitrification
(C) Nitrogen fixation
(D) Nitrification

39. **Nitrification is inhibited under which condition**
(A) Anaerobic and Highly acidic soil
(B) Anaerobic and Highly alkaline soil
(C) Aerobic and Highly acidic soil
(D) Aerobic and highly alkaline soil

40. **Addition of material high in protein such as sewage and manure will stimulate process of**
(A) Nitrogen fixation
(B) Denitrification
(C) Ammon*ifi*cation
(D) Nitrification

41. **The product of nitrification which is toxic to human beings is**
(A) Nitrite
(B) Nitrate
(C) Nitrous oxide
(D) Nitrogen dioxide

42. **Nitrite combines with haemoglobin to produce toxic compound**
(A) Ninhemoglobin
(B) Nithemoglobin
(C) Methemoglobin
(D) Methylohemoglobin

43. **Permeable limit of nitrate in water is**
(A) 10 ppm (B) 20 ppm
(C) 30 ppm (D) 100 ppm

44. **Nitrification inhibitor is**
(A) 2-chloro 6- (trichloromethyle) pyridine
(B) Nitropyrin
(C) Both
(D) None

45. **Stem nodulation in Sesbania is done by**
(A) *Azorhizobium caulinodans*
(B) *Bradyrhizobium japonicum*
(C) *Rhizobium trifoli*
(D) *Rhizobium leguminoserum*

46. **Which chemical can increase efficiency of ammonical fertilizers by inhibiting conversion of ammonia to nitrate**
(A) Nitrification inhibitor
(B) Denitrification inhibitor
(C) Ammon*ifi*cation inhibitor
(D) Nitrogen fixation inhibitors

47. **Symbiotic nitrogen fixing bacteria are**
(A) *Rhizobium*
(B) *Bradyrhizobium*
(C) *Azorhizobium*
(D) All

48. ***Rhizobium* fix atmospheric nitrogen in symbiotic association with**
(A) Cereal plants
(B) Leguminous plants
(C) Deciduous plants
(D) All

49. ***Rhizobium* invades legume roots from**
(A) Primary root (B) Stem hairs
(C) Root hairs (D) Stem base

50. **Host specificity is observed in which N fixing bacteria**
(A) *Rhizobium* (B) *Acetobacter*
(C) *Azotobacter* (D) *Azospirillum*

51. **Soybean plants infected by**
(A) *Rhizobium meliloti*
(B) *Bradyrhizobium japonicum*
(C) *Rhizobium leguminoserum*
(D) *Rhizobium trifoli*

52. ***Rhizobium meliloti* infects which of the following plants**
(A) Alfalfa (B) Soybean
(C) Pea (D) Clover

53. **Clover plants nodules are formed by**
(A) *R. leguminoserum*
(B) *R. Meliloti*
(C) *R. trifoli*
(D) *R. japonicum*

54. **In pea, green gram, black gram the *Rhizobium* found is**
(A) *R. leguminoserum*
(B) *R. japonicum*
(C) *R. trifoli*
(D) *R. meliloti*

55. **In Bengal gram which *Rhizobium* is found**
(A) *Azorhizobium*
(B) *Rhizobium*
(C) *Bradyrhizobium*
(D) *Synorhizobium*

56. **Nitrogen is an important component of**
(A) Alkaloids (B) Amides
(C) Amino acids (D) All of above

57. **Anaerobic nitrogen fixing bacteria is**
(A) *Clostridium pasteurianum*
(B) *Clostridium botulinm*
(C) *Clostridium perfringens*
(D) *Clostridium pneumonia*

58. **Anaerobic nitrogen fixing bacteria was first isolated by**
(A) Frank (B) Beijerinck
(C) Winogradsky (D) Hiltner

59. **Who first isolated aerobic nitrogen fixing bacteria *Azotobacter chroococcum***
(A) M. Beijerinck
(B) S. Winogradsky
(C) A. B. Frank
(D) L. Pasteur

60. **Nitrogenase enzyme is made up of**
(A) Fe protein (B) Mo-Fe protein
(C) Both (D) None

61. ***Nif* cluster is made up of**
(A) 17-20 *nif* genes
(B) 18-25 *nif* genes
(C) 10-12 *nif* genes
(D) 8-10 *nif* genes

62. **In nitrogen fixation how many molecules of ammonia are formed from one molecule of nitrogen gas**
(A) 1 (B) 2
(C) 4 (D) 5

63. **How much ATP are utilized for nitrogen fixation**
(A) 16 (B) 18
(C) 12 (D) 14

64. **How much hydrogen and electron molecules are required for N fixation**
(A) 4 (B) 8
(C) 6 (D) 2

65. **Electron donor for N fixation is**
(A) Ferrodoxin (B) Pyruvate
(C) Oxygen (D) Acetyl Co A

66. **Pyruvate donates its electrons to**
(A) Ferrodoxin (B) Pyruvate
(C) Oxygen (D) Ferrocyanin

67. **Beans are infected by**
(A) *R. phaseoli*
(B) *R. leguminoserum*
(C) *R. japonicum*
(D) *R. trifoli*

68. **Fe protein in nitrogenase enzyme is encoded by**
(A) *nif* D gene (B) *nif* K gene
(C) *nif* H gene (D) *nif* Y gene

69. **Mo- Fe protein is encoded by**
(A) *nif* D gene and *nif* K gene
(B) *nif* H gene and *nif* L
(C) *nif* H and *nif* D
(D) *nif* K and *nif* L

70. **Regulatory protein in *nif* cluster are**
(A) *nif* L and *nif* A
(B) *nif* D gene and *nif* K gene
(C) *nif* H gene and *nif* L
(D) *nif* H and *nif* D

71. Synthesis of *nif* structural genes such as *nif* H D K is started by

(A) *nif* L (B) *nif* A
(C) *nif* J (D) *nif* M

72. *nif* A is which type of regulator for *nif* expression

(A) Positive (B) Negative
(C) It is not regulator
(D) It do not regulate *Nif* genes

73. factor involved in *nif* gene expression and synthesis is

(A) σ 28 (B) σ 51
(C) σ 54 (D) σ 64

74. Fe-protein is an enzyme

(A) Nitrogenase reductase
(B) Nitrogenase
(C) Nitrate reductase
(D) Nitrite reductase

75. In nitrogen fixation how many ATPs are required for fixation of 1 molecule of N_2

(A) 6 (B) 8
(C) 3 (D) 4

76. Most energy expensive process in nature is

(A) Nitrogen fixation
(B) Nitrate reduction
(C) Nitrogen assimilation
(D) Denitrification

77. Molecular weight of Fe-protein is

(A) 220 KD (B) 60 KD
(C) 120 KD (D) 421 KD

78. Molecular weight of Mo-Fe protein is

(A) 230 KD (B) 320 KD
(C) 220 KD (D) 160 KD

79. Flavonoids secreted by leguminous plants are

(A) Naringenin and Daidzein
(B) Genistein and Glycitein
(C) Naringenin and Genistein
(D) Daidzein and Glycitein

80. The first nod gene which recognize flavonoid of plant is

(A) Nod D (B) Nod E
(C) Nod A (D) Nod M

81. Nod D gene activates synthesis of

(A) Nod A (B) Nod B
(C) Nod C (D) All

82. Bacteroids are formed within which part of the roots

(A) Xylem (B) Phloem
(C) Cortex (D) Vessels

83. Host specificity observed in *Rhizobium* is due to which nod factor

(A) Nod D (B) Nod H
(C) Nod A (D) Nod E

84. Important enzymes involved in nitrogen fixation are

(A) Nitrogenase and hydrogenase
(B) Nitrogenase and hexokinase
(C) Nitrogenase and peptidase
(D) Nitrogenase and hydrolyase

85. The chief source of nitrogen for green plants is

(A) Atmospheric nitrogen
(B) Nitrates
(C) Ammonium salts
(D) Low molecular weight- organic nitrogenous compound

86. In the presence of carbon monoxide, nitrogen fixation

(A) Increases (B) Accelerated
(C) Inhibits (D) Not affected

87. The enzyme dinitrogenase is

(A) Monomer (B) Dimer
(C) Trimer (D) Tetramer

88. Root nodules contains pink pigment having high affinity for oxygen is

(A) Haemoglonin
(B) Leghaemoglobin
(C) Bacterial haemoglobin
(D) Nod haemoglobin

89. **Which of the following genus of bacteria comprise of free living nitrogen fixers**
(A) *Azotobacter* (B) *Azospirillum*
(C) *Rhizobium* (D) *Acetobacter*

90. **Which of the following group of organisms maintains high respiration rate for protecting nitrogenase from oxygen deactivation**
(A) *Azotobacter* (B) *Azospirillum*
(C) *Rhizobium* (D) *Acetobacter*

91. **Which of the following group of organisms comprise of associative nitrogen fixers and generally found in association with cereals**
(A) *Azotobacter* (B) *Azospirillum*
(C) *Rhizobium* (D) *Acetobacter*

92. **Which of following genus of microorganisms are microaerophilic in nature**
(A) *Azotobacter* (B) *Azospirillum*
(C) *Rhizobium* (D) *Acetobacter*

93. **Regulatory element in *nif* cluster is**
(A) *nif* L (B) *nif* A
(C) *nif* H (D) *nif* D

| Answer key |

Q. No.	Ans	Q. No.	Ans	Q. No.	Ans	Q. No.	Ans	Q. No.	Ans	Q. No.	Ans
1.	C	2.	A	3.	C	4.	C	5.	A	6.	B
7.	B	8.	B	9.	A	10.	A	11.	A	12.	D
13.	C	14.	C	15.	B	16.	B	17.	A	18.	D
19.	D	20.	A	21.	D	22.	C	23.	B	24.	D
25.	A	26.	A	27.	C	28.	D	29.	A	30.	A
31.	B	32.	A	33.	C	34.	B	35.	A	36.	C
37.	C	38.	D	39.	A	40.	D	41.	A	42.	C
43.	A	44.	C	45.	A	46.	A	47.	D	48.	B
49.	C	50.	A	51.	B	52.	A	53.	C	54.	A
55.	C	56.	D	57.	A	58.	C	59.	A	60.	C
61.	A	62.	B	63.	A	64.	B	65.	B	66.	A
67.	A	68.	C	69.	A	70.	A	71.	B	72.	A
73.	C	74.	A	75.	D	76.	A	77.	B	78.	C
79.	A	80.	A	81.	D	82.	C	83.	B	84.	A
85.	B	86.	C	87.	D	88.	B	89.	A	90.	A
91.	B	92.	B	93.	A						

◈

Chapter **14**

MICROBIAL TRANSFORMATIONS OF PHOSPHORUS, SULPHUR AND MINOR CARBON NUTRIENTS

1. **The process in which simple organic compounds are decomposed by microorganism to release simple inorganic compounds is known as**
(A) Dissimilation (B) Assimilation
(C) Immobilization (D) Mineralization

2. **The process of incorporation of a simple, soluble substance into the body of an organism, is known as**
(A) Immobilization (B) Assimilation
(C) Mineralization (D) Dissimilation

3. **Reduction of sulphate to sulphide under anaerobic condition is the process known as**
(A) Assimilatory sulphate reduction
(B) Dissimilatory sulphate reduction
(C) Assimilatory suphide reduction
(D) Dissimilatory suphide reduction

4. **Dissimilatory suphate reduction occurs under**
(A) Aerobic condition
(B) Microaerophilic condition
(C) Anaerobic condition
(D) Oxygenic condition

5. **Bacteria which carry out dissimilatory sulphate reduction are**
(A) Desulfovibrio desulfuromonas
(B) *Alteromonas* and *Clostridium*
(C) *Desulfovibrio* and *Desulfotomaculum*
(D) All of above

6. **Which of the following is sedimentary cycle**
(A) Oxygen cycle
(B) Carbon cycle
(C) Phosphorous cycle
(D) Nitrogen cycle

7. **In phosphorous cycle, phosphorous becomes available by weathering of rocks first to**
(A) Consumer
(B) Decomposer
(C) Primary Producer
(D) Secondary Producer

8. **In dissimilatory sulphate reduction, the external electron donor is**
(A) Sulphate
(B) Sulphite
(C) Elemental Sulphur
(D) Gaseous Sulphur

9. **Which microorganisms reduce sulphite to sulphide**
(A) Alteromonas and *Clostridium*
(B) Desulfovibrio and *Desulfotomaculum*
(C) *Alteromonas*
(D) *Clostridium*

10. **Which of the following are aerobic sulphur oxidizing bacteria?**
(A) *Beggiatoa* (B) *Thiothrix*
(C) *Alteromonas* (D) *Clostridium*

11. Which of the following are Photolithotrophic sulphur oxidizing bacteria

(A) *Chromatium* and *Chlorobium*
(B) *Desulfovibrio* and *Desulfotomaculum*
(C) *Alteromonas* and *Clostridium*
(D) *Chromatium* and *Clostridium*

12. Which of the following are anaerobic sulphur oxidizing bacteria?

(A) *Chromatium* and *Chlorobium*
(B) *Desulfovibrio* and *Desulfotomaculum*
(C) *Thiobacillus* and *Beggiatoa*
(D) All of above

13. Which of the following bacteria converts H_2S in to elemental sulphur?

(A) *Thiobacillus*
(B) *Gallionella*
(C) *Sulfolobus*
(D) All of above

14. Transformation of ferrous ion (Fe^{+2}) in to ferric ion (Fe^{+3}) is carried out by ....

(A) Iron oxidizing bacteria
(B) Iron reducing bacteria
(C) Iron assimilating bacteria
(D) Iron dissimilating bacteria

15. Iron oxidizing bacteria active under acidic condition is

(A) *Thiobacillus ferrooxidans*
(B) *Gallionella*
(C) *Desulfotomaculum*
(D) *Chlorobium*

16. Which of the following is an Iron oxidizing bacteria active under neutral pH?

(A) *Gallionella* (B) *Sulfolobus*
(C) *Chlorobium* (D) *Thiobacillus*

17. Which of the following bacteria carry out iron oxidation under acidic and thermophilic condition?

(A) *Sulfolobus*
(B) *Chromatium*
(C) *Thiobacillus ferrooxidans*
(D) *Thiobacillus thiooxidans*

18. Which of the following are Iron reducing bacteria?

(A) *Geobacter metallireducens* and *Geobacter sulfurreducens*
(B) *Ferribacterium limneticum* and *Shewanella putrefaciens*
(C) *Desulfuromonas acetoxidans* and *Pelobacter carbinolicus*
(D) All of above

19. In acidic soil insoluble form of phosphorous is ....

(A) Calcium phosphate
(B) Aluminium phosphate
(C) Iron phosphate
(D) Magnesium phosphate

20. The key mechanism of inorganic phosphate solubilization is

(A) Organic acid production
(B) Siderophore production
(C) Mineralization
(D) Phosphatase production

21. Which of the following are phosphate solubilizing microorganisms?

(A) *Pseudomonas*
(B) *Bacillus*
(C) *Aspergillus*
(D) All of above

22. Organic form of phosphorous is

(A) Nucleic acid (B) Lecithine
(C) Phytic acid (D) All of above

23. The process in which organic form of phosphorous is converted in to inorganic form is known as ....

(A) Mineralization
(B) Solubilization
(C) Stabilization
(D) All of above

24. Removal of contaminants from natural habitat using microorganisms is known as ....
(A) Bioremediation
(B) Biodegradation
(C) Biotransformation
(D) Bioreconstruction

25. The Bioremediation process in which indigenous bacteria are stimulated by supply of air and nutrients through wells is known as
(A) Bioventing
(B) Biosparging
(C) Bioshaking
(D) Bioairlifting

26. Which one the following is most appropriate bioremediation strategy to clean up deep under surface site contaminated with Hydrocarbon or oil
(A) Bioventing
(B) Biosparging
(C) Rhizofiltration
(D) Rhizoremediation

27. Bioremediation process involving supply of oxygen and nutrients by circulating aqueous solutions through contaminated soils is known as ....
(A) Bioventing
(B) Biosparging
(C) Rhizofiltration
(D) Rhizoremediation

28. The law which states that total biomass of an organism will be determined by the nutrient present in the lowest concentration relative to the organism's requirements.
(A) Leibig's law of minimal
(B) Shelford's law of tolerance
(C) Koch's postulates
(D) Pasteur effect

29. The law which states that there are limits to environmental factors below and above which a microorganism cannot survive and grow, regardless of the nutrient supply .
(A) Shelford's law of tolerance
(B) Leibig's law of minimal
(C) Koch's postulates
(D) Pasteur effect

30. Degradation or Transformation of a substance in to new compounds through biochemical reactions or the actions of microorganisms
(A) Bioremediation
(B) Phytoremediation
(C) Biodegradation
(D) Phytodegradation

31. The first super bug having capacity to degrade Camphor, octanol, xylene, naphthalene
(A) *Agrobacterium tumefacience*
(B) *Bacillus megaterium*
(C) *Pseudomonas fluorescence*
(D) *Pseudomonas putida*

32. The most radio resistant bacteria are
(A) *Pseudomonas putida*
(B) *Deinococcus radiodurans*
(C) *Bacillus megaterium*
(D) *Pseudomons radiodurans*

33. The process in which plant take up organic compounds and convert it in to more stable less toxic and less mobile form is known as
(A) Phytotransformation
(B) Phytoextraction
(C) Phytostabilization
(D) All of above

34. The process in which the pollutant become stable mass of plant is known as
(A) Phytotransformation
(B) Phytoextraction
(C) Phytostabilization
(D) Phytomobilization

35. The degradation of contaminant by rhizospheric bacteria is known as ...

(A) Rhizofiltration
(B) Rhizodegradation
(C) Rhizostabilization
(D) Rhizotransformation

36. Which of the following is water bioremediation technique?

(A) Rhizofiltration
(B) Rhizostabilization
(C) Phytodegradation
(D) Rhizostabilisation

37. Complex, brown, amorphous and colloidal organic substance formed by microbial decomposition is known as

(A) Soil (B) Sand
(C) Humus (D) Clay

38. Humic group of compounds are extracted from humus by which method?

(A) Acid extraction
(B) Solvent extraction
(C) Carbon Dioxide extraction
(D) Alkali extraction

39. Humic group of compounds are made up of

(A) Polyketones (B) Polyphenols
(C) Polyhexones (D) Polypropanes

40. Humic group of compounds make up about how much % of soil organic matter

(A) 50-60 % (B) 20-30 %
(C) 60-80 % (D) 10-20 %

41. Humic group of compounds are ....to microbial attack

(A) Resistant
(B) Susceptible
(C) Highly Susceptible
(D) None of above

42. Non humic group of compounds make up how much % of humus

(A) 10-20 % (B) 40-50 %
(C) 20-30 % (D) 60-80 %

43. Which of the following are known as non humic group of substances?

(A) Polysaccharides
(B) Polyuronides proteins,
(C) Tannins
(D) All of above

44. This theory states that humic substances are formed due to the incomplete degradation of lignin

(A) Lignin theory
(B) Humin theory
(C) Humic acid theory
(D) Humus theory

45. This theory states that humic substances are formed by cellulose decomposing mycobacteria earlier to lignin decomposition

(A) Kononovas theory
(B) Flaig theory
(C) Sochtig theory
(D) Waksman theory

46. This theory states that the humic substances are formed by the condensation of phenolic materials

(A) Humin theory
(B) Polyphenol theory
(C) Humic acid theory
(D) Lignin theory

47. Humus is composed of which elements

(A) Carbon
(B) Hydrogen
(C) Oxygen
(D) All of above

48. Humus contain .... charge

(A) Negative
(B) Positive
(C) None
(D) Both

49. Fulvic acid contains how many minerals
(A) 90 or more
(B) 40 or more
(C) 70 or more
(D) 10 or more

50. Which component of humus is good antioxidant
(A) Humin
(B) Fulvic acid
(C) Humic acid
(D) Flavin

51. Which humus substance is used for transport of minerals to cells
(A) Humin
(B) Fulvic acid
(C) Humic acid
(D) Humonuric acid

52. Which humus substance is known as hidden treasure
(A) Humic acid
(B) Fulvic acid
(C) Humin
(D) Flavin

| Answer key |

Q. No.	Ans	Q. No.	Ans	Q. No.	Ans	Q. No.	Ans	Q. No.	Ans	Q. No.	Ans
1.	D	2.	A	3.	B	4.	C	5.	D	6.	C
7.	B	8.	A	9.	C	10.	A	11.	C	12.	C
13.	A	14.	A	15.	A	16.	A	17.	A	18.	D
19.	A	20.	A	21.	D	22.	D	23.	A	24.	A
25.	A	26.	A	27.	B	28.	A	29.	A	30.	C
31.	D	32.	B	33.	A	34.	C	35.	B	36.	A
37.	C	38.	D	39.	B	40.	C	41.	A	42.	C
43.	D	44.	A	45.	A	46.	B	47.	D	48.	A
49.	C	50.	B	51.	B	52.	B				

Chapter 15

BASIC MICROBIOLOGICAL TECHNIQUES AND ENVIRONMENTAL MICROBIOLOGY

1. **A population of cells arising from a single cell is known as ....**
   (A) Pure culture (B) Mixed culture
   (C) Single culture (D) Mono Culture

2. **Which of the following scientist has developed pure culture techniques?**
   (A) Robert Koch (B) Louis Pasteur
   (C) Richard Petri (D) Joseph Lister

3. **Which of the following technique is utilized for isolation and culturing of sample containing small number of microorganisms?**
   (A) Enrichment culture method
   (B) Diluted culture technique
   (C) Pour plate technique
   (D) Spread plate technique

4. **Which of the following is an advanced technique for isolation of single microbial cell from complex environment?**
   (A) Micromanipulator and Optical tweezers
   (B) Enrichment culture technique
   (C) Standard plate method
   (D) Electron microscopy

5. **Which of the following is common method of preservation of microorganisms?**
   (A) Periodic transfer
   (B) Lyophilization
   (C) Freezing
   (D) Oil overlaying

6. **Which is the main limitation of periodic transfer?**
   (A) Mutation
   (B) Phenotypic change
   (C) Loss of traits
   (D) All of above

7. **Which of the following technique is the best for preservation of spore forming microorganisms**
   (A) Soil preservation
   (B) Oil overlaying
   (C) Freezing
   (D) Periodic transfer

8. **In ultra-freezing, microorganisms are preserved in liquid nitrogen at which temperature?**
   (A) -196 °C (B) -200 °C
   (C) -180 °C (D) -250 °C

9. **Which of the following process removes all living cells, viable spores, viruses, and viroids?**
   (A) Sterilization (B) Disinfection
   (C) Immobilization (D) Pasteurization

10. **By which of the following process disease causing microorganisms can be removed but spores can't?**
    (A) Sterilization (B) Lyophilisation
    (C) Disinfection (D) Pasteurization

11. **The process in which the microbial population is reduced to levels that are considered safe by public health standards is known as ....**

(A) Sterilization (B) Disinfection
(C) Sanitization (D) Antisepsis

12. **Chemical agents applied to tissue to prevent infection by killing or inhibiting pathogen growth are known as**
(A) Antibiotics (B) Sanitizers
(C) Sterilants (D) Antiseptics

13. **Which of the following represents the lowest temperature at which a microbial suspension is killed in 10 minutes**
(A) Thermal death point
(B) D value
(C) Z value
(D) F value

14. **Which of the following represents the time required to kill 90% of the microorganisms or spores in a sample at a specified temperature?**
(A) Thermal death point
(B) D value
(C) Z value
(D) F value

15. **Which of the following represents increase in temperature required to reduce D value to 1/10?**
(A) Thermal death point
(B) D value
(C) Z value
(D) F value

16. **Which of the following represents the time in minutes at a specific temperature (usually 250°F or 121°C) needed to kill a population of cells or spores**
(A) Thermal death point
(B) D value
(C) Z value
(D) F value

17. **Which of the following represents shortest time needed to kill all organisms in a microbial suspension at a specific temperature and under defined conditions?**
(A) Thermal death time
(B) Thermal death point
(C) Z value
(D) F value

18. **In an autoclave, microorganisms get killed due to which of the following reason?**
(A) Denaturation of proteins
(B) Disruption of plasma membrane
(C) Degradation of nucleic acid
(D) All of above

19. **How much time is required to sterilize 5 litre of liquid at 121°C temperature in an autoclave?**
(A) 20 min (B) 30 min
(C) 50 min (D) 70 min

20. **Which of the following is biological indicator for sterilization?**
(A) *Bacillus stearothermophilus*
(B) *Clostridium* PA3679
(C) Both
(D) None

21. **In low temperature high time (LTHT) method the milk is held under which of the following conditions?**
(A) 63°C for 30 minutes
(B) 72°C for 15 seconds
(C) 140 to 150°C for 1 to 3 seconds
(D) None of above

22. **In high temperature short time method (HTST) or flash sterilization method the object is held under which of the following conditions?**
(A) 63°C for 30 minutes
(B) 72°C for 15 seconds
(C) 140 to 150°C for 1 to 3 seconds
(D) None of above

23. Which instrument is used in dry heat sterilization

(A) Hot air oven (B) Autoclave
(C) Boiling (D) Pasteurization

24. Which of the following events occur during dry heat sterilization to kill microorganisms?

(A) Oxidation of cell constituents
(B) Denaturation of proteins
(C) Denaturation of enzymes
(D) All of above

25. Heat sensitive solutions can be sterilized by which of the following method

(A) Filtration (B) Boiling
(C) Pasteurization (D) Sterilization

26. Which of the following microorganisms cannot be removed by membrane filtrations

(A) Virus (B) Bacteria
(C) Fungal spore (D) Protozoa

27. Efficiency of HEPA filter is or laminar air flow is .... %

(A) 100 (B) 99.97
(C) 90.97 (D) 95.97

28. Which of the following ionizing radiation are used for sterilization?

(A) UV radiation
(B) Gamma Radiation
(C) Both
(D) None

29. Which of the following method of sterilization is also known as cold sterilization

(A) Radiation (B) Filtration
(C) Autoclaving (D) Pasteurization

30. Which of the following is first widely used antiseptic and disinfectant?

(A) Phenol (B) Ethanol
(C) Methanol (D) Propanol

31. Which of the following compound is commercial disinfectant comprising of mixture of phenolic compounds?

(A) Lysol (B) Tysol
(C) Xylitol (D) Resorcinol

32. Which of the following phenolic compound is most popular antiseptic?

(A) Hexachlorophene
(B) Tetrachlorophene
(C) Heptachlorophene
(D) Trichlorophene

33. How much concentration of alcohol is used for disinfection?

(A) 90-95% (B) 50-60%
(C) 70-80% (D) 40-50%

34. How much time will it take to obtain disinfection using alcohol?

(A) 10-15 min (B) 5-10 min
(C) 15-20 min (D) 15-30 min

35. Which of the following is commonly used disinfectant for municipal water supplies?

(A) Iodine (B) Fluorine
(C) Chlorine (D) Bromine

36. Which of the following is gaseous sterilant or disinfectant?

(A) Chlorine (B) Fluorine
(C) Bromine (D) Iodine

37. Which of the following chlorinated compound used as disinfectant?

(A) Chlorine gas
(B) Sodium hypochlorite
(C) Calcium hypochlorite
(D) All of above

38. How much time it requires to disinfect object with chlorine?

(A) 30 minutes
(B) 10 minutes
(C) 20 minutes
(D) 60 minutes

39. **Which of the following compound can be considered as replacement of chlorine for disinfection of drinking water?**
(A) Ozone
(B) Carbon dioxide
(C) Fluorine (D) Bromine

40. **Which of the following compound is lethal to microorganism ?**
(A) Mercuric bromide
(B) Mercuric sulphate
(C) Mercuric chloride
(D) Mercuric nitrate

41. **Which compound is algicide used in lakes and swimming pools?**
(A) Copper sulphate
(B) Iron sulphate
(C) Zinc sulphate
(D) Aluminium sulphate

42. **Which of the following is also known as detergent?**
(A) Ammonium sulphate
(B) Quaternary ammonium compounds
(C) Fatty acid
(D) Ammonium chloride

43. **Which of the following chemical agents combine with microbial nucleic acid?**
(A) Halogen (B) Alcohol
(C) Aldehydes (D) Mercury

44. **Which of the following aldehyde containing sterilant that can kill microbial spores?**
(A) Formaldehyde
(B) Glutaraldehyde
(C) Both (D) None

45. **Which gas sterilant can enter packaging material even plastic wraps?**
(A) Ethylene oxide
(B) Chlorine (C) Formaldehyde
(D) Iso-propyl alcohol

46. **Quality or potency of disinfectant is judged by which of the following measure?**
(A) Phenol coefficient
(B) Chlorine coefficient
(C) Sulphur coefficient
(D) Oxygen coefficient

47. **The phenomenon of bending of light rays when light passes from one medium to another is known as ....**
(A) Resolution
(B) Magnification
(C) Refraction
(D) Numerical aperture

48. **Which of the following phenomenon of microscope states that the image should remain in focus when objectives are changed?**
(A) Resolution
(B) Par focalization
(C) Refraction
(D) Magnification

49. **Which of the following phenomenon defines capacity of lens to show separate image of two closely placed objects?**
(A) Resolution
(B) Par focalization
(C) Refraction
(D) Magnification

50. **Refractive index of air and glass is .... and .... respectively.**
(A) 1.5 and 1.0 (B) 1.0 and 2.0
(C) 0.5 and 1.0 (D) 1.0 and 1.5

51. **Resolving power of normal microscope is ....**
(A) 0.2 nm (B) 0.2 μm
(C) 0.2 mm (D) 0.2 cm

52. **In fluorescent microscope, which type of light source is used?**
(A) Ultraviolet (B) Violet
(C) Blue (D) All of above

53. **Which of the following is source of light in fluorescent microscope**
(A) Mercury lamp
(B) Cobalt lamp
(C) Halogen lamp
(D) UV lamp

54. **Which of the following is most critical step of Gram's staining which brings about differentiation?**
(A) Primary staining
(B) Decolorization
(C) Secondary staining
(D) Mordent application

55. **Which of the following compound is responsible for acidfastness of Mycobacterium?**
(A) Mycolic acid (B) Teichoic acid
(C) Muramic acid (D) All of above

56. **As the wavelength of light decreases Resolution ....**
(A) Decreases (B) Increases
(C) Not affected (D) First increase and then decrease

57. **Wavelength of electron is**
(A) 0.005 cm (B) 0.005 μm
(C) 0.005 km (D) 0.005 nm

58. **Resolving power of electron microscope is ....**
(A) 0.5 μm
(B) 5 Å
(C) 0.5 cm
(D) 5 m

59. **Which of the following is light source for transmission electron microscope?**
(A) Tungsten filament lamp
(B) Mercury lamp
(C) Halogen lamp
(D) Cobalt 60 lamp

60. **In electron microscop,e electrons pass through which medium?**
(A) Glass (B) Air
(C) Vaccum (D) All of above

61. **Which of the following chemical is used as fixative in TEM?**
(A) Glutaraldehyde and Osmium tetroxide
(B) Gelatine and Glutaraldehyde
(C) Galacturoladehyde and Formaldehyde
(D) Osmium trioxide and Gelatine

62. **Which of the following chemical is used to dehydrate specimen in TEM?**
(A) Chlorine (B) Acetone
(C) Ammonia (D) Ethanol

63. **Which of the following microscopy reveals internal cellular structures or organelles?**
(A) Transmission Electron Microscopy
(B) Scanning Electron Microscopy
(C) Fluorescent Microscopy
(D) Dark Field Microscopy

64. **Which of the following light is suitable for getting maximum resolution**
(A) Red (B) Blue
(C) Orange (D) Green

65. **The electrons of the Scanning Electron Microscope are reflected through**
(A) Glass funnel (B) Specimen
(C) Metal-coated surfaces
(D) Vacuum chamber

66. **The object can be magnified under an electron microscope about**
(A) 350,000 times (B) 250,000 times
(C) 300,000 times (D) 450,000 times

67. **Which of the following microscopy technique is utilized for observing live cells**
(A) TEM (B) SEM
(C) Phase contrast microscopy
(D) Fluorescent microscopy

68. Which of the following is known as an air pollution indicator?

(A) Lichen (B) *Aeromonas*
(C) *Aerobacillus* (D) *Aerococcus*

69. Which of the following is water pollution indicator?

(A) *Oscillatoria* (B) *Chlorella*
(C) *Spirulina* (D) Lichens

70. Which of the following is an index indicating harmful level of pollutant in air?

(A) Gas quality index
(B) Oxygen quality index
(C) SMOG index
(D) Air quality index

71. Which one of the following indicates range of air quality index?

(A) 0-50 (B) 0-500
(C) 0-1000 (D) 0-1500

72. Partial burning of carbon in the fuel result in production of which colorless gas?

(A) Carbon monoxide
(B) Carbon dioxide
(C) Hydrogen sulphide
(D) Chlorofluoro carbon

73. Which of the following is reddish brown gas responsible for air pollution?

(A) Nitrous oxide (B) Ozone
(C) Nitrogen dioxide
(D) Ferrous oxide

74. Combination of gases with water vapour and dust is known as ....

(A) DMOG (B) SMOG
(C) VMOG (D) LMOG

75. Which of the following air pollutant is produced by volcanoes

(A) Sulfur dioxide
(B) Carbon monoxide
(C) Carbon dioxide
(D) Ozone

76. Which of the following gas is responsible for acid rain?

(A) Carbon monoxide
(B) Carbon dioxide
(C) Ozone
(D) Sulfur dioxide

77. Which one of the following is yellow green gas pollutant?

(A) Bromine
(B) Chlorine
(C) Iodine
(D) Nitrogen dioxide

78. Half life of DDT in soil is .... years

(A) 10 (B) 11
(C) 13 (D) 15

79. Zone of partially decomposed organic matter and inorganic compounds is known as ....

(A) Z horizon (B) A horizon
(C) B horizon (D) C horizon

80. Which soil horizon is known as subsoil?

(A) Z horizon (B) A horizon
(C) B horizon (D) C horizon

81. Which soil horizon contains broken rocks?

(A) Z horizon
(B) A horizon
(C) B horizon
(D) C horizon

82. Which soil horizon contains un-weathered rocks?

(A) R horizon
(B) A horizon
(C) B horizon
(D) C horizon

83. Single definable source of water pollution is known as ....

(A) Point source
(B) Non-point source
(C) Specific source
(D) Non-specific source

84. Which of the following is non-point source of water pollution?

(A) Run off from farmlands
(B) Animal feed lots
(C) Rural areas
(D) All of above

85. If the water body is rich in nutrients such as nitrogen and phosphorous which type of microorganisms will proliferate?

(A) Bacteria
(B) Algae
(C) Fungi
(D) Protozoa

86. Which of the following phenomenon represents process of concentrating hazardous chemicals in to food chain?

(A) Biomagnification
(B) Biointensification
(C) Biomultiplication
(D) Biostimulation

87. Which of the following curve defines the microbial activity and oxygen demand of point source contaminated water?

(A) Oxygen sag curve
(B) BOD curve
(C) Oxygen demand curve
(D) Microbial oxygen curve

88. How many zones are present in an oxygen sag curve?

(A) 2 (B) 4
(C) 6 (D) 8

89. Among water pollution, which type of pollution is hard to be removed ?

(A) Ground water pollution
(B) Domestic water pollution
(C) Surface water pollution
(D) Sea water pollution

90. Presumptive test starts with the calculation of microorganisms by which technique?

(A) Colony count
(B) Direct microscopic count
(C) Most probable number (MPN) count
(D) All of above

91. Which tests are performed for detection of coliform organisms in water body?

(A) Presence- absence test (P-A) test
(B) Defined substrate test (DST)
(C) Membrane filtration techniques
(D) All of above

92. For evaluating microbiological properties of waste water which method is most preferable?

(A) Presence- absence test (P-A) test
(B) Defined substrate test (DST)
(C) Membrane filtration techniques
(D) All of above

93. For simultaneous testing of water samples for the presence of coliform and *E. coli* which test is used

(A) Related colilert defined substrate test
(B) Related duality defined substrate test
(C) Related collaborative defined substrate test
(D) Related *Enterobacter* defined substrate test

94. Which of the following is tested for determination of water quality

(A) Organic carbon content
(B) Organic nitrogen content
(C) Available nitrogen content
(D) Available phosphorous content

95. Organic carbon content is measured by

(A) Total organic carbon (TOC)
(B) Chemical oxygen demand (COD)
(C) Biochemical oxygen demand (BOD)
(D) All of above

96. Chemically oxidizable carbon is detected by ....

(A) Chemical oxygen demand
(B) Biochemical oxygen demand
(C) Biological oxygen demand
(D) All of above

97. In COD which compound is not detected?

(A) Lignin (B) Cellulose
(C) Hemicellulose (D) All of above

98. Which of the following is an indirect measure of organic matter in aquatic environment?

(A) COD (B) BOD
(C) OM (D) OC

99. The amount of dissolved O_2 needed for microbial oxidation of biodegradable organic matter is known as ....

(A) COD (B) BOD
(C) OM (D) OC

100. Which of the following compound/s is/are added to BOD test?

(A) Nitrification inhibitors
(B) 2-chloro-6-(trichloromethyl) pyridine
(C) Nitrapyrin
(D) All of above

101. How much time will it take to complete BOD test?

(A) 2 days (B) 3 days
(C) 5 days (D) 10 days

102. Primary treatment of waste water can remove ..... % BOD

(A) 10-20 (B) 20-30
(C) 30-40 (D) 40-50

103. In which type of waste water treatment, microbiological removal of dissolved organic matter occurs?

(A) Primary (B) Secondary
(C) Tertiary (D) Quaternary

104. How much BOD is removed by secondary treatment?

(A) 70-80 % (B) 80-90 %
(C) 90-95% (D) 90-100%

105. Which of the following is aerobic secondary waste water treatment?

(A) Activated sludge
(B) Trickling filter
(C) Extended aeration
(D) All of above

106. In which of the following secondary waste water treatment, suspended microbes are utilized to remove organic matter?

(A) Activated sludge
(B) Trickling filter
(C) Extended aeration
(D) Aerated lagoons

107. In which of the following waste water treatment, non-biodegradable organic material, heavy metals and minerals are removed?

(A) Primary (B) Secondary
(C) Tertiary (D) Quaternary

108. Permissible value of chlorine residues in purified water is ....

(A) 0.2-2.0 mg/l
(B) 0.02-0.2 mg/l
(C) 0.02-0.04 mg/l
(D) 0.1-1.0 mg/l

109. Which of the following disinfectant by product is formed in water purification system?

(A) Trihalomethane
(B) Trihaloethane
(C) Trihalopropane
(D) Trihalobutane

110. Ideal compost is having C: N ratio ....

(A) < 30 (B) >30
(C) < 20 (D) >20

111. If N content of substrate is high, than decomposition is........

(A) Slower (B) Faster

(C) Not affected (D) Not depended

112. Protein rich materials are decomposed ....

(A) Faster (B) Slower

(C) Not affected (D) Not depended

113. Wide C:N ratio results in ....

(A) Slower decomposition

(B) Faster decomposition

(C) Not affected

(D) Not dependent

114. How much quantity of methane is present in biogas?

(A) 50-60 % (B) 30-40 %

(C) 10-20 % (D) 80-90 %

115. In which method of disposal of municipal solid waste, the waste is dumped in the soil?

(A) Incineration (B) Composting

(C) Land filing (D) Shredding

116. The process of decomposition of biodegradable solid waste by earthworms is called

(A) Land fills

(B) Shredding

(C) Vermi-composting

(D) Composting

117. Which of the following is not the land filling method?

(A) Bangalore method

(B) Area method

(C) Depression method

(D) Trench method

118. The liquid that passes through solid waste and extracts suspended impurities from it.

(A) Leachate

(B) Sludge

(C) Distilled water

(D) Municipal waste

119. In which method of composting, decomposition of anaerobic waste takes place?

(A) Indian method

(B) Depression method

(C) Bangalore method

(D) Trench method

120. Which of the following gas can be produced from landfill wastes?

(A) Biogas

(B) Oxygen gas

(C) Liquified petroleum gas

(D) Butane

121. Which of the following gas generates when the matter inside the sanitary landfill breaks down?

(A) Methane (B) Nitrogen

(C) Hydrogen (D) Oxygen

122. The process of burning of municipal solid waste at high temperature is called

(A) Composting (B) Land filing

(C) Incineration (D) Shredding

123. Methanogenic phase in biogas production is

(A) Strict anaerobic

(B) Strict aerobic

(C) Microaerophilic

(D) Facultative anaerobic

| Answer key |

Q. No.	Ans	Q. No.	Ans	Q. No.	Ans	Q. No.	Ans	Q. No.	Ans	Q. No.	Ans
1.	A	2.	A	3.	A	4.	A	5.	A	6.	D
7.	A	8.	A	9.	A	10.	C	11.	C	12.	D
13.	A	14.	B	15.	C	16.	D	17.	A	18.	D
19.	A	20.	C	21.	A	22.	B	23.	A	24.	D
25.	A	26.	A	27.	B	28.	C	29.	A	30.	A
31.	A	32.	A	33.	C	34.	B	35.	C	36.	A
37.	D	38.	A	39.	A	40.	C	41.	A	42.	B
43.	C	44.	C	45.	A	46.	A	47.	C	48.	B
49.	A	50.	D	51.	B	52.	D	53.	A	54.	B
55.	A	56.	B	57.	D	58.	B	59.	A	60.	C
61.	A	62.	B	63.	A	64.	B	65.	C	66.	C
67.	C	68.	A	69.	A	70.	D	71.	B	72.	A
73.	C	74.	B	75.	A	76.	D	77.	B	78.	D
79.	A	80.	C	81.	D	82.	A	83.	A	84.	D
85.	B	86.	A	87.	A	88.	B	89.	A	90.	C
91.	D	92.	C	93.	A	94.	A	95.	D	96.	A
97.	A	98.	B	99.	B	100.	D	101.	C	102.	B
103.	B	104.	C	105.	D	106.	A	107.	C	108.	A
109.	A	110.	C	111.	B	112.	A	113.	A	114.	A
115.	C	116.	C	117.	A	118.	A	119.	C	120.	A
121.	A	122.	C	123.	A						

◈

Chapter 16

MICROBIAL BIOTECHNOLOGY

1. **During citric acid fermentation which of the following metal ions are added to stop growth of *Aspergillus niger***
   (A) Magnesium and iron
   (B) Manganese and iron
   (C) Iron and copper
   (D) Copper and molybdenum

2. **In citric acid fermentation how much concentration of sugar is used?**
   (A) 10-15 % (B) 20-22 %
   (C) 20-25 % (D) 15-18 %

3. **Which of the following metal ion counteract inhibition of citric acid production by iron?**
   (A) 0.2 ppm Copper
   (B) 0.02 ppm Nickel
   (C) 0.2 ppm Magnesium
   (D) 0.1 ppm Manganese

4. **Which of the following enzyme is involved in production of gluconic acid?**
   (A) Gluconate synthase
   (B) Glucose oxidase
   (C) Galactose Oxidase
   (D) Gluconate dehydrogenase

5. **Which of the following microorganism is used for gluconic acid production?**
   (A) *Aspergillus niger*
   (B) *Aspergillus flavus*
   (C) *Lactobacillus plantarum*
   (D) *Sachharomyces cerevicea*

6. **Which of the following is the key ingredient of the medium for gluconic acid production?**
   (A) Molasses
   (B) Coconut water
   (C) Corn steep liquor
   (D) Corn starch

7. **Which of the following microorganism is used for production of acetic acid?**
   (A) *Acetobacter* spp.
   (B) *Lactobacillus* spp.
   (C) *Aspergillus* spp.
   (D) *Nostoc* spp.

8. **Which of the following microorganism is used for production of citric acid**
   (A) *Aspergillus flavus*
   (B) *Lactobacillus plantarum*
   (C) *Aspergillus niger*
   (D) *Sachharomyces cereviceae*

9. **Which of the following medium is used for production citric acid?**
   (A) Corn steep liquor based medium
   (B) Coconut water based medium
   (C) Corn starch based medium
   (D) Molasses based medium

10. **Which of the following microorganism is used for fumaric acid production?**
    (A) *Rhizopus stolonifer*
    (B) *Rhizopus oryzae*
    (C) *Rhizopus nigricans*
    (D) *Rhizopus americanus*

11. **Which of the following microorganism is used for production of itaconic acid?**
(A) *Aspergillus terreus*
(B) *Rhizopus nigricans*
(C) *Aspergillus niger*
(D) *Rhizopus stolonifer*

12. **Which of the following technique is used for sterilization of must?**
(A) Pasteurization and boiling
(B) Pasteurization and treatment of ethylene oxide
(C) Pasteurization and phenol treatment
(D) Pasteurization and treatment of sulphur dioxide

13. **The science of wine production is known as ....**
(A) Winology (B) Ecology
(C) Enology (D) Oncology

14. **Which of the following organisms can be used for wine production?**
(A) *Sachharomyces cereviceae* and *Sachharomyces ellipsoideus*
(B) *Saccharomyces uvarum* and *Sachharomyces cereviceae*
(C) *Saccharomyces pastorianus* and *Sachharomyces cereviceae*
(D) *Saccharomyces pastorianus* and *Sachharomyces ellipsoideus*

15. **How much concentration of alcohol is present in wine?**
(A) 5-10 % (B) 8-10 %
(C) 10-18 % (D) 20-25 %

16. **Which of the following procedure is used to remove sediments produced by microbial growth during wine production?**
(A) Racking
(B) Kasking
(C) Pasteurization
(D) All of above

17. **Which of the following is distilled wine?**
(A) Rum (B) Whisky
(C) Beer (D) Brandy

18. **Sparkling wine is known as**
(A) Rum (B) Whisky
(C) Beer (D) Champagnes

19. **Which of the following microorganisms used for production of beers?**
(A) *Sachharomyces carlsbergensis*
(B) *Sachharomyces cereviceae*
(C) *Sachharomyces delbrueckii*
(D) *Sachharomyces ellipsoideus*

20. **Which of the following microorganisms used for production of Ale?**
(A) *Sachharomyces carlsbergensis*
(B) *Sachharomyces cereviceae*
(C) *Sachharomyces delbrueckii*
(D) *Sachharomyces ellipsoideus*

21. **Which of the following beverage is distilled beer?**
(A) Whiskey (B) Brandy
(C) Rum (D) Champagnes

22. **How much percentage of rye grain is present in rye beer?**
(A) 40 % (B) 51 %
(C) 70 % (D) 20 %

23. **Which of the following is known as peripheral equipment in fermenter?**
(A) Sparger
(B) Baffles
(C) Impellers
(D) Buffer bottles

24. **In which of the following type of fermenter impellers are used?**
(A) Stirred tank fermenter
(B) Air lift fermenter
(C) Solid state fermenter
(D) Fluidized bed reactor

25. In which of the following fermenter air is used to mix medium?

(A) Stirred tank fermenter
(B) Air lift fermenter
(C) Solid state fermenter
(D) Fluidized bed reactor

26. In which of the following fermenter microorganism as biofilm are held on the surface of the solid support material particles which are tightly held and the fermentation medium passes through the support material?

(A) Stirred tank fermenter
(B) Air lift fermenter
(C) Solid state fermenter
(D) Fixed bed reactor

27. In which of the following fermenter support particles are suspended in medium on to which microorganisms are attached as biofilm?

(A) Stirred tank fermenter
(B) Air lift fermenter
(C) Solid state fermenter
(D) Fluidized bed reactor

28. Which of the following are primary metabolite?

(A) Amino acids, nucleotides, ethanol and lactic acid
(B) Pigments, alkaloids, toxins and antibiotics
(C) Gibberellins, carotenoids, toxins and antibiotics
(D) Ethanol, lactic acid, pigments and alkaloids

29. Which of the following are secondary metabolites?

(A) Antibiotics, Mycotoxins and Plant growth hormones
(B) Amino acid, Lactic acid, Ethanol
(C) Amino acid, Nucleotide, Ethanol
(D) Lactic acid, Ethanol, Mycotoxins

30. Primary metabolites accumulate during which growth phase?

(A) Lag phase
(B) Log phase
(C) Stationary phase
(D) Decline phase

31. Secondary metabolites accumulate during which growth phase?

(A) Lag phase
(B) Log phase
(C) Stationary phase
(D) Decline phase

32. For growing filamentous bacteria or fungi, which type of fermenter is most suitable?

(A) Stirred tank fermenter
(B) Air lift fermenter
(C) Fluidized bed reactor
(D) Fixed bed reactor

33. In which of the following method, pollutants are removed using plants?

(A) Bioremediation
(B) Bidegradation
(C) Phytoremediation
(D) None of above

34. Which of the following is an intrinsic factor associated with food spoilage?

(A) Moisture content
(B) Relative humidity
(C) Temperature
(D) Presence of gases

35. Which of the following is an extrinsic factor associated with food spoilage?

(A) Presence of gases
(B) Water activity
(C) Oxidation reduction potential
(D) Structure of food

36. Which of the following process leads to foul smelling in meat and butter?

(A) Putrefaction (B) Pectinolysis
(C) Lipolysis (D) Ransidation

37. Which of the following microorganism can readily spoil carbohydrate rich food?

(A) Bacteria (B) Viruse
(C) Mycoplasma (D) Fungi

38. Which of the following process leads to soft rot in fruits?

(A) Putrefaction (B) Pectinolysis
(C) Lipolysis (D) Proteolysis

39. Which of the following is an anti-microbial substance present in eggs?

(A) Lysozyme (B) Allicin
(C) Eugenol (D) Nisin

40. Garlic contains which antimicrobial compound?

(A) Lysozyme (B) Allicin
(C) Eugenol (D) Nisin

41. Clove contains which antimicrobial compound?

(A) Lysozyme (B) Allicin
(C) Eugenol (D) Polyphenol

42. Which of the following antimicrobial compound is present in green and black tea?

(A) Lysozyme (B) Allicin
(C) Eugenol (D) Polyphenol

43. Modified atmospheric packaging (MAP) uses how much CO_2?

(A) 30 % (B) 40 %
(C) 50 % (D) 60 %

44. Which of the following fungus is responsible for ergotism in grains?

(A) *Claviceps purpurea*
(B) *Aspergillus niger*
(C) *Aspergillus flavus*
(D) *Fusarium moniliforme*

45. Which of the following fungi produces Aflatoxin?

(A) *Claviceps purpurea*
(B) *Aspergillus niger*
(C) *Aspergillus flavus*
(D) *Fusarium moniliforme*

46. Which of the following fungi produces monisin toxin?

(A) *Claviceps purpura*
(B) *Aspergillus niger*
(C) *Aspergillus flavus*
(D) *Fusarium moniliforme*

47. In appertization process the canned food is heated in retort at .... °C

(A) 121 (B) 115
(C) 120 (D) 110

48. Sulphur stinker is caused by production of which gas?

(A) Hydrogen sulphide
(B) Hydrogen cyanide
(C) Hydrogen peroxide
(D) Hydrogen

49. Sulphur stinker is produced in which type of food?

(A) Highly acidic
(B) Highly alkaline
(C) Neutral
(D) Alkaline

50. Breads contain which preservative?

(A) Sodium benzoate
(B) Sodium chloride
(C) Citric acid
(D) Sodium propionate

51. Which of the following chemical compound is used as preservative for meat items?

(A) Sodium propionate
(B) Sodium nitrite
(C) Sodium benzoate
(D) Citric acid

52. Which of the following radiation is generally used for food preservation?

(A) X-rays
(B) Gamma rays
(C) UV rays
(D) Infrared rays

53. How much quantity of gamma irradiation is used to preserve meat?

(A) 4.5 to 5.6 Mrd (B) 1.2 to 2.4 Mrd

(C) 2.3 to 4.6 Mrd (D) 5.2 to 6.4 Mrd

54. Which of the following microorganism produce bacteriocin nisin?

(A) *Streptococcus lactis*

(B) *Lactobacillus lactis*

(C) *Lactobacillus acidophilus*

(D) *Streptococcus thermophilus*

55. Starchy foods are mainly spoiled by which organisms?

(A) *Clostridium perfringens*

(B) *Clostridium botulinum*

(C) *Bacillus cereus*

(D) *Campylobacter jejuni*

56. Majority of food borne illness caused worldwide are mainly due to

(A) *Clostridium perfringens*

(B) *Clostridium botulinum*

(C) *Bacillus cereus*

(D) *Campylobacter jejuni*

57. Which of the following microorganisms are used as starter culture for production of yogurt?

(A) *S. thermophilus* and *L. bulgaricus* in 1:1 ratio

(B) *S. thermophilus* and *L. bulgaricus* in 1:2 ratio

(C) *L. bulgaricus* and *S. thermophilus* in 1:3 ratio

(D) *L. bulgaricus* and *S. thermophilus* in 2:1 ratio

58. Which of the following microorganisms are used as probiotic bacteria?

(A) *Lactobacillus acidophilus* and *Bifidobacterium*

(B) *Lactobacillus bulgaricus* and *Streptococcus thermophiles*

(C) *Lactobacillus acidophilus* and *Lactobacillus bulgaricus*

(D) *Streptococcus thermophilus* and *Bifidobacterium*

59. Which of the following probiotic bacteria may produce Vitamin B complex and increase absorption of calcium?

(A) *Lactobacillus acidophilus*

(B) *Bifidobacterium*

(C) *Lactobacillus bulgaricus*

(D) *Streptococcus thermophiles*

60. Which of the following fermented milk product contains 2 % alcohol?

(A) Yogurt (B) Curd

(C) Kefir (D) Villi

61. Which of the following fermented milk product is formed by fermentation of milk with *Geotrichium candidum*?

(A) Yogurt (B) Curd

(C) Kefir (D) Villi

62. How much amount of lactic acid is present in villi?

(A) 0.5 % (B) 0.9 %

(C) 1.2 % (D) 1.5 %

63. Rennet is used in

(A) Bread making

(B) Cheese making

(C) Fermentation

(D) Antibiotic synthesis

64. Which of the following cheese contains eye formation?

(A) Swiss cheese

(B) Camembert (blue) cheese

(C) Roquefort cheese

(D) Mozzarella cheese

65. Which microorganism contributes in final flavour development and eye formation in swiss cheese

(A) Propionibacterium

(B) Lactobacillus

(C) Geotrichium

(D) Pediococcus

66. Which cheese is soaked in brine to stimulate the development of specific fungi and bacteria

(A) Limburger (B) Camembert

(C) Swiss (D) Roquefort

67. Which of the following is fermented meat product

(A) Sufu (B) Silage

(C) Tofu (D) Sausage

68. Which of the following organisms are desired for sausage production?

(A) *Sachharomyces cerevisiae and Lactobacillus plantarum*

(B) *Pediococcus cerevisiae* and *Lactobacillus plantarum*

(C) *Geotrichium candidum and Pediococcus cerevisiae*

(D) *Sachharomyces cerevisiae and Geotrichium candidum*

69. Which of the following is chemically coagulated soybean milk product?

(A) Tofu (B) Kefir

(C) Izuki (D) Silage

70. Which of the following microorganisms used for production of Tempeh from soybean mash?

(A) *Lactobacillus* sp

(B) *Penicillium* sp.

(C) *Pediococcus* sp.

(D) *Rhizopus* sp

71. Which of the following is fermented cabbage?

(A) Sausage (B) Silage

(C) Sauerkraut (D) Izuki

72. Salt (NaCl) concentration in sauerkraut is

(A) 1.2-2.2 % (B) 2.2-2.8 %

(C) 3.0-3.2 % (D) 3.2-4.0 %

73. Which of the following is fermented food product containing cucumber?

(A) Pickle (B) Izuki

(C) Sauerkraut (D) Silage

74. Which of the following fermented milk product contains alcohol?

(A) Kefir (B) Tofu

(C) Sufu (D) Izuki

75. Which of the following is known as fermented feed for animal?

(A) Sausage (B) Kefir

(C) Silage (D) Izuki

76. Which of the following yeast can be used as single cell protein?

(A) *Candida utilis*

(B) *Candida lipolytica*

(C) *Saccharomyces cerevisiae*

(D) All of above

77. A single cell protein refers to

(A) A specific protein extracted from pure culture of single type of cell

(B) Sources of mixed proteins extracted from pure and mixed culture of microorganisms or cells

(C) Protein extracted from a single cell

(D) A specific protein extracted from a single cell

78. Which of the following fungi is used as SCP?

(A) *Aspergillus fumigatus*

(B) *Aspergillus niger*

(C) *Rhizopus cyclopean*

(D) All of above

79. Which of the following bacteria is used as SCP?

(A) *Pseudomonas fluorescens*

(B) *Lactobacillus acidophilus*

(C) *Bacillus megaterium*

(D) All of above

80. Spirulina is

(A) Edible fungi

(B) Biofertilizer

(C) Biopesticide

(D) Single cell protein

81. Which of the following algae is used as SCP?

(A) *Chlorella stigmatophora*

(B) *Scenedesmus acutus*

(C) *Spirulina maxima*

(D) All of above

82. Fermented high moisture stored fodder is known as

(A) Silage (B) Sludge

(C) Sufu (D) Pickel

83. Which of the following is the major product on molar basis in silage

(A) Lactic acid (B) Acetic acid

(C) Ethanol (D) Carbon dioxide

84. Which cleanup approach involves removing groundwater or soil from its natural setting to allow for bioremediation?

(A) In situ bioremediation

(B) Ex situ bioremediation

(C) Bioaugmentation

(D) Phytoremediation

85. Which bioremediation approach involves mixing contaminated soil with water, carbon dioxide and fertilizers in a bioreactor to stimulate biodegradation?

(A) In situ hybridization

(B) Slurry-phase bioremediation

(C) Biopile treatment

(D) In situ bioremediation

86. Bioaugmentation is a process that involves:

(A) Using plants for bioremediation

(B) Bioventing

(C) Sludge removal

(D) Adding microbes to a cleanup site

87. The process in which the halophytes (plants adapted to saline soil) to extract salt from the soil to improve its fertility

(A) Phytodesalination

(B) Rhizofiltration

(C) Phytovolatilization

(D) Phytostimulation

88. The phytoremediation process which involves enhancement of soil microbial activity for the degradation of organic contaminants, typically by organisms that associate with roots

(A) Phytodesalination

(B) Rhizofiltration

(C) Phytovolatilization

(D) Phytostimulation

89. The plant assisted process which reduces the mobility of substances in the environment by binding the contaminant in soil near to root region is known as

(A) Phytostabilization

(B) Rhizofiltration

(C) Phytoextraction

(D) Phytostimulation

90. The process in which plant remove contaminants from soil or water into harvestable plant biomass is known as

(A) Phytostabilization

(B) Rhizofiltration

(C) Phytoextraction

(D) Phytostimulation

91. The process in which removal of substances from soil or water with release into the air occurs is known as

(A) Phytodesalination

(B) Rhizofiltration

(C) Phytovolatilization

(D) Phytostimulation

92. Which of the following tree plants are more promising for phytoremediation

(A) Willow

(B) Poplar

(C) Alpine

(D) All of above

93. Frozen storage is generally operated at temperature of:

(A) -0°C (B) -18°C
(C) -50°C (D) -60°C

94. Jam, jellies can be preserved by adding sugar at concentration of:

(A) 65% (B) 75%
(C) 40% (D) 30%

95. Tocopherol is an example of:

(A) Anticaking agent
(B) Flavouring agent
(C) Antioxidant
(D) Antitoxin

96. Mold inhibitor used in bread is:

(A) Calcium carbonate
(B) Sodium chloride
(C) Sodium / Calcium propionate
(D) Sodium benzoate

97. Who developed the process of canning?

(A) Nicolas Appert
(B) Louis Pasteur
(C) Norman Borlaug
(D) Walter Hesse

98. The first synthetic sweetening agent used was:

(A) Saccharin (B) Cyclamates
(C) Aspartame (D) Sucralose

99. Nisin is used as:

(A) Antimicrobial agent
(B) Emulsifier
(C) Stabilizer
(D) Sweetner

100. The Limiting Amino acid in green vegetables is:

(A) Arginine
(B) Lysine
(C) Methionine
(D) Tryptophan

101. Butylated Hydroxyanisole (BHA) is:

(A) Chelating agent
(B) Antioxidant
(C) Stablizer
(D) Emulsion

102. Which Vitamin is not found in Fruits and Vegetables

(A) Vitamin A
(B) Vitamin B1
(C) Vitamin B6
D) Vitamin B12

103. Caesium-137 and strontium-90 were removed from a pond using which of the following plant after the Chernobyl accident

(A) Sunflowers (B) Tomato
(C) Alpine (D) Willow

| Answer key |

Q. No.	Ans	Q. No.	Ans	Q. No.	Ans	Q. No.	Ans	Q. No.	Ans	Q. No.	Ans
1.	B	2.	D	3.	A	4.	B	5.	A	6.	C
7.	A	8.	C	9.	D	10.	C	11.	A	12.	D
13.	C	14.	A	15.	C	16.	A	17.	D	18.	D
19.	A	20.	B	21.	A	22.	B	23.	D	24.	A
25.	B	26.	D	27.	D	28.	A	29.	A	30.	B
31.	C	32.	B	33.	C	34.	A	35.	A	36.	D
37.	A	38.	B	39.	A	40.	B	41.	C	42.	D
43.	D	44.	A	45.	C	46.	D	47.	B	48.	A
49.	A	50.	D	51.	B	52.	B	53.	A	54.	A
55.	C	56.	A	57.	B	58.	A	59.	B	60.	C
61.	D	62.	B	63.	B	64.	A	65.	A	66.	A
67.	D	68.	B	69.	A	70.	D	71.	C	72.	B
73.	A	74.	A	75.	C	76.	D	77.	B	78.	D
79.	D	80.	D	81.	D	82.	A	83.	A	84.	B
85.	B	86.	D	87.	A	88.	D	89.	A	90.	C
91.	C	92.	D	93.	B	94.	A	95.	C	96.	C
97.	A	98.	A	99.	A	100.	C	101.	B	102.	D
103.	A										

◈